Styliani Avraamidou, Efstratios Pistikopoulos
Multi-level Mixed-Integer Optimization

Also of Interest

Process Intensification
Breakthrough in Design, Industrial Innovation Practices, and Education
Jan Harmsen, Maarten Verkerk, 2020
ISBN 978-3-11-065734-0, e-ISBN (PDF) 978-3-11-065735-7,
e-ISBN (EPUB) 978-3-11-065752-4

Product-Driven Process Design
From Molecule to Enterprise
Edwin Zondervan, Cristhian Almeida-Rivera, Kyle Vincent Camarda,
2020
ISBN 978-3-11-057011-3, e-ISBN (PDF) 978-3-11-057013-7,
e-ISBN (EPUB) 978-3-11-057019-9

Process Intensification
Design Methodologies
Edited by Fernando Israel Gómez-Castro, Juan Gabriel
Segovia-Hernández, 2019
ISBN 978-3-11-059607-6, e-ISBN (PDF) 978-3-11-059612-0,
e-ISBN (EPUB) 978-3-11-059279-5

Integrated Chemical Processes in Liquid Multiphase Systems
From Chemical Reaction to Process Design and Operation
Edited by Matthias Kraume, Sabine Enders, Anja Drews,
Reinhard Schomäcker, Sebastian Engell, Kai Sundmacher, 2022
ISBN 978-3-11-070943-8, e-ISBN (PDF) 978-3-11-070985-8,
e-ISBN (EPUB) 978-3-11-070991-9

Styliani Avraamidou, Efstratios Pistikopoulos

Multi-level Mixed-Integer Optimization

—

Parametric Programming Approach

DE GRUYTER

Mathematics Subject Classification 2010
Primary: 90-00, 90C11, 90C90, 91A05, 91A06; Secondary: 90C29, 90C31, 90C47

Authors

Prof. Styliani Avraamidou
University of Wisconsin-Madison
Dept. of Chemical and Biological Engineering
1415 Engineering Dr
Madison
WI 53706
USA
avraamidou@wisc.edu

Prof. Dr. Efstratios Pistikopoulos
Texas A & M University
Dept. of Chemical Engineering
327 Giesecke Engineering Research Building
College Station
TX 77843-3251
USA
stratos@tamu.edu

ISBN 978-3-11-076030-9
e-ISBN (PDF) 978-3-11-076031-6
e-ISBN (EPUB) 978-3-11-076038-5

Library of Congress Control Number: 2021950449

Bibliographic information published by the Deutsche Nationalbibliothek
The Deutsche Nationalbibliothek lists this publication in the Deutsche Nationalbibliografie;
detailed bibliographic data are available on the Internet at http://dnb.dnb.de.

© 2022 Walter de Gruyter GmbH, Berlin/Boston
Cover image: natrot / iStock / Getty Images Plus
Typesetting: VTeX UAB, Lithuania
Printing and binding: CPI books GmbH, Leck

www.degruyter.com

Preface

Multilevel optimization formulations can be applied for the solution of a broad range of decision making problems across different fields, including economics, sciences and engineering. Multilevel optimization problems are a class of optimization problems that involves a set of nested optimization problems over a single feasible region. The control over the decision variables is divided among different optimization levels, but all decision variables can affect the objective function and constraints of all optimization levels.

Multilevel optimization problems are very challenging to solve even when considering just two linear decision levels. For classes of problems where the lower level problems also involve discrete variables, the difficulty is further increased, typically requiring global optimization methods for its solution. Theoretic and algorithmic developments on the solution of mixed-integer multilevel optimization problems, along with many applications in the areas of economics and process systems engineering, have been constantly emerging during the last 20 years. Our group has published over 20 manuscripts on the subject of multi-level optimization solution algorithms through multiparametric programming. We have also developed a MATLAB® based toolbox, B-POP®, for the solution of various classes of multilevel optimization problems.

This book aims to enable fundamental understanding in the area of mixed-integer multilevel optimization. More specifically, this book focuses on the solution of (i) bilevel mixed-integer linear programming problems, (ii) bilevel mixed-integer quadratic programming problems, (iii) trilevel mixed-integer linear and quadratic programming problems, (iv) bilevel multifollower mixed-integer linear and quadratic programming problems and (v) multilevel nonlinear programming problems. We hope that by the end of this book, the reader will be able to not only understand how to formulate multilevel optimization problems, but also be able to solve, both by hand and through computational tools continuous and mixed-integer multilevel optimization problems.

The book begins with a motivation on why to formulate decision making problems as multilevel optimization problems, with different application examples for different classes of multilevel optimization problems. Then a background on multi-level optimization follows, summarizing the key contributions and key approaches of solution methods in the field. The first chapter finishes with an overview of multiparametric programing, summarizing over key concepts that will be used in the following chapters of the book.

The second chapter focuses on bilevel mixed-integer optimization. An algorithm based on multiparametric programming is first introduced to solve bilevel mixed-integer linear optimization problems and then expanded for the solution of bilevel mixed-integer quadratic optimization problems and bilevel problems with right-hand side uncertainty. All three algorithms are explained step-by-step through multiple numerical examples.

https://doi.org/10.1515/9783110760316-201

The third chapter generalizes and expands the algorithms presented in the second chapter, to address first the solution of tri-level linear and quadratic optimization problems, and then more general mutlilevel optimization problems, including nonlinear and multifollower problems. Similar to the second chapter, all algorithms presented in this chapter are explained step-by-step through numerical examples.

The fourth chapter includes a variety of application case studies, from industrial engineering, chemical engineering and operational research, including classical bilevel problems such as the integration of production and distribution planning, and other novel applications such as a hierarchical economic model predictive controller and a class of robust optimization. The formulation and solution of these problems is presented in detail.

Finally, the last chapter consider es the implementation of the algorithms discussed in Chapters 2 and 3 in a MATLAB® based toolbox, B-POP®, along with computational studies to highlight the capabilities of the algorithms.

This book is the outcome of research work carried out at the Center for Process Systems Engineering of Imperial College London and the Texas A&M Energy Institute of Texas A&M University. We would like to take the opportunity to thank former and current PhD students, and post-doctorate/research associates from our research team that have been involved in the presented work, particularly Dr. Nikolaos A. Diangelakis and Dr. Richard Oberdieck.

We would also like to gratefully acknowledge the financial support kindly provided by our many sponsors: EPSRC, NSF, EU/ERC, DOE/CESMII, DOE/RAPID and Shell. Finally, we would like to thank De Gruyter for their enthusiastic support for this book.

Styliani Avraamidou
Efstratios N. Pistikopoulos

Contents

1 Introduction

In this chapter, a motivation on why to formulate decision making problems as multilevel optimization problems is presented. Different application examples for different classes of multilevel optimization problems are discussed. Then a background on multilevel optimization follows, summarizing the key contributions and key approaches for the solution of multilevel optimization problems. This first chapter finishes with an overview of multiparametric programming, summarizing the key concepts that will be used in the following chapters of the book.

1.1 Motivation

In the real world, many decision making processes involve different decision makers that lie in a hierarchy. Decisions are made at different levels of the hierarchy and can affect the outcome of all decision makers. For such cases, we incline to focus on just one level of the decision making and take the other levels into consideration as assumptions [1]. Taking into consideration all decision levels as optimization problems is referred to as multilevel optimization. This field has gained a lot of attention through the years and become a well-known and significant research field [2–4].

Multilevel optimization and decision making can be applied in many and diverse disciplines such as bioengineering, chemical and civil engineering, mechanics, management, network design, transportation and economics while new applications are constantly being proposed. The rest of this chapter will introduce some motivating applications for specific classes of multilevel mixed-integer optimization problems that will be tackled in the later chapters of this book.

1.1.1 Bilevel programming

Optimization problems involving two decision makers at two different decision levels are referred to as bilevel programming problems. Bilevel programming has attracted the most attention among other classes of multilevel programming problems due to its simplicity (compared to other multilevel problems) and great applicability. It has been applied to many and diverse problems that require hierarchical decision making such as transportation network planning [1, 5, 6], urban planning [7], economic planning [8–10], management [11], design under uncertainty [12–14], design and control integration [15–18], supply chain planning [19–21] and parameter estimation [22].

In this book, we will focus on solution methods for mixed-integer bilevel problems, and its application on the integration of production and distribution planning (Section 4.1).

https://doi.org/10.1515/9783110760316-001

Application A: Production and distribution planning
Supply chains are systems with multiple decision levels corresponding to different activities, spanning from the procurement of raw materials to the distribution of the final products to the costumers. Even though these decisions are interlinked and can affect each other, in most cases they are considered individually [20, 23]. The significance of the integration of production and distribution decisions inside supply chains, in order to account for the interactions between them, has been recognized by different researchers [20, 24, 25]. Proposed integrated approaches include assuming (i) that one company controls the integrated process by owning both the processing plants and distribution centers [26–28], or (ii) that the processing plants and distribution centers are owned by different companies, each trying to optimize their own objective [19, 21, 29–31], as presented in figure 1.1.

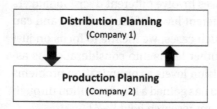

Figure 1.1: Schematic representation of the production-distribution planning problem with different companies controlling the distribution and production of products.

Considering the second case, the production-distribution planning (PD) problem can be expressed as a hierarchical decision making problem, involving two different decision makers corresponding to each company. Assuming one company owns the production plants and another the distribution centers, the resulting problem is a two-level hierarchical decision making problem. The first level is responsible for optimizing the distribution centers overall costs and is influenced by the second level that is responsible for optimizing the production plants overall costs.

When considering the PD problem, decisions taken at both decision levels can involve both continuous (e. g., production rates, distribution rates) or discrete (e. g., choice of production plant, choice of distribution center, active routes) variables. Therefore, the integrated PD problem results into a mixed-integer bilevel programming problem.

An indicative list of publications focusing on the problem of distribution and production planning integration is presented in Table 1.1.

1.1.2 Bilevel programming under uncertainty

For applications that involve constantly changing and unpredictable conditions, it is of high importance to consider the effect of uncertainties in programming problems.

Table 1.1: Indicative list of publications on distribution and production planning integration.

Reference	Contribution
Erenguc et al. [24]	Integrated production/distribution planning in supply chains
Gupta and Maranas [26]	Two-stage modeling for production-distribution systems
Jung et al. [28]	Supply chain management under demand uncertainty
Jaber and Osman [32]	Two-level (supplier-retailer) supply chain coordination
Sousa et al. [27]	Supply Chain Design and multilevel planning
Kuo and Han [30]	Production-Distribution bilevel model and solution method
Calvete et al. [19]	Production-Distribution bilevel model and solution method

When considering bilevel programming formulations, uncertainties can be both integer or continuous, and can arise in both optimization levels. In this book, we will discuss two different applications of this class of problems: (i) *supply chain planning*: unstable business environment, with constantly changing market conditions and customer needs and expectations [26, 28, 33], and (ii) *hierarchical model predictive control*: constantly changing system states and unpredicted system disturbances [34].

Application B: Supply chain planning under uncertainty

Sources of uncertainty in supply chain planning may include variations in processing rates, canceled or rushed orders, equipment failure, raw material, final product or utility price fluctuations and demand variations [35].

Considering the PD problem introduced earlier in this section (Application A), a key source of uncertainty is the product demand. Failure to consider this in PD planning could lead to either unsatisfied customer demands and loss of market share, or excessively high inventory holding costs [36]. Since the PD is a bilevel mixed-integer problem, taking into consideration demand uncertainty would result into a bilevel mixed-integer problem with right-hand side uncertainty.

Application C: Hierarchical model predictive control

Hierarchical control structures consist of a hierarchy of control levels, where every level is controlling a subset of the overall control variables, by manipulating a subset of the overall manipulated variables [37–39]. In the case of hierarchical model predictive control (MPC) structures, each control level involves one or more optimization problem, with the resulting formulation typically corresponding to a multilevel programming problem with a single second level problem (referred to as follower) (Figure 1.2(a)) or multiple second level problems (Figure 1.2(b)).

When attempting to solve hierarchical control problems using bilevel programming, one would need to solve a bilevel problem at each control time step. Since bilevel problems are very challenging to solve (see Section 1.2.1), the computational time required to solve such problems could be very big, therefore, it would only be possible to apply such a controller in applications with very slow dynamics.

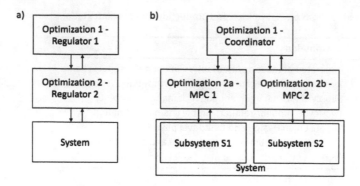

Figure 1.2: Hierarchical Process Control Systems: (a) two-level control system with a single first and second control levels, (b) two-level control system with a single first and two lower control levels.

A way to overcome this challenge would be to use explicit bilevel controllers and, therefore, solve the bilevel problem under uncertainty once and offline. More details on this application area can be found in Section 4.2.

1.1.3 Trilevel programming

Optimization problems involving three decision makers at three different decision levels are referred to as trilevel programming problems. This class of problems has attracted considerable attention across a broad range of communities, including economics, operation research, engineering and management. It can be applied to many and diverse problems that require three-level hierarchical decision making such as safety and defense [40–42], supply chain management [43], energy management [44] and robust optimization [45–47]. In this book, we will explore the application of trilevel optimization for the solution of a class of Adjustable Robust Optimization (ARO) problems (Section 4.3).

Application D: Adjustable robust optimization
One of the dominant approaches to address decision making under uncertainty is robust optimization. Adjustable Robust Optimization (ARO) problems are an extended class of classical robust optimization problems that include two types of variables, "here-and-now" variables that are to be decided before the realization of the uncertainty, and "wait-and-see" variables that are to be decided after the realization of the uncertainty. In Section 4.3, we formulate this problem as a tri-level optimization problem where the first optimization level is controlling the "here-and-now" variables, the second optimization level is maximizing over the uncertainty and the third optimization level is controlling the adjustable "wait-and-see" variables, as illustrated in Figure 1.3.

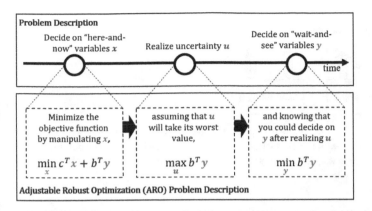

Figure 1.3: Schematic representation of ARO as a trilevel problem with the first optimization level minimizing the objective by manipulating the "here and now" decisions, the second optimization level maximizing the objective by manipulating choosing the worst uncertainty for the chosen "here and now" variables, and in turn, the third optimization level minimizing the objective by manipulating the "wait and see" decision variables.

1.1.4 Bilevel multifollower programming

Optimization problems involving two optimization levels with the second level containing multiple optimization problems (followers) are referred to as bilevel multifollower programming problems. Bilevel multifollower programming (MF-B) can be applied to many and diverse problems such as transportation network planning [1, 5, 6], urban planning and land-use planning [7], economic planning [8–10], design and control integration [15–18, 48], and supply chain planning [19–21, 48]. In this book, we will focus on two applications, (i) a planning and scheduling integration problem, and (ii) a design and operation integration problem.

Application E: Planning and scheduling integration
Traditionally, process planning and scheduling strategies are derived sequentially and separately. Scheduling decisions are derived after process planning decisions are already taken. This can lead to suboptimal strategies, therefore, researchers have tried to integrate these decision levels and solve this problem holistically [36, 49, 50].

Planning and scheduling optimization problems with seasonal demand variability can be often expressed holistically within a hierarchical structure, where optimal decisions at an upper level (planning) provide constraints for the detailed decision making (scheduling) at a lower level, typically posed as bilevel multi-follower optimization problems [14, 51–53] (Figure 1.4). Since discrete decisions are involved most likely at both levels, the resulting formulations typically correspond to bilevel multi-follower mixed integer linear programming (MF-B-MILP) problems. An inticative list of research papers on the integration of planning and scheduling is presented in Table 1.2.

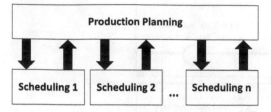

Figure 1.4: Schematic representation of the planning and scheduling integration problem as a bilevel multifollower problem, with the production planning problem as the first-level optimization problem, and many scheduling problems inside the time period of the planning problem as the second-level followers.

Table 1.2: Indicative list of publications that describe the planning and scheduling integration problem as a hierarchical multifollower optimization problem.

Reference	Contribution
Gershwin [54]	Hierarchical scheduling and planning in manufacturing systems
Petkov and Maranas [36]	Multiperiod planning and scheduling under demand uncertainty
Bose and Pekny [55]	Model predictive framework for planning and scheduling
Pinto et al. [56]	Planning and scheduling models for refinery operations
Lee et al. [57]	Planning and scheduling with outsourcing in supply chain
Moon et al. [50]	Integrated process planning and scheduling
Kim et al. [49]	Integration of process planning and job shop scheduling
Li and Ierapetritou [53]	Production planning and scheduling integration

Application F: Design and operation integration

To derive optimal designs for economical and efficient plants, integration of the design and operation decision making processes might be needed. Design decisions, such as plant and unit capacity, choice of raw materials, products and processing steps, need to be decided while taking into account the operating decisions that can include scheduling decisions and production targets.

In this book, we will focus on the integration of design and scheduling, where optimal decisions at an upper level (design) provide constraints for the detailed decision making (scheduling) at a lower level, as illustared in figure 1.5. The integrated problem is posed as bilevel multifollower optimization problem. The mathematical formulation and more details regarding this application area can be found in Section 4.5.

1.2 A background on multilevel optimization

1.2.1 Bilevel continuous optimization

Optimization problems involving two decision makers at two different decision levels are referred to as bilevel programming problems: the first decision maker (upper level; leader) is solving an optimization problem, which includes in its constraint

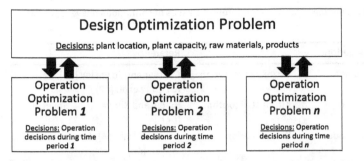

Figure 1.5: Schematic representation of the design and operation integration problem as a bilevel multifollower problem, with the design problem as the first-level optimization problem, and many operation optimization problems inside the time period of the design problem as the second-level followers.

set another optimization problem solved by the second decision maker (lower level; follower).

Bilevel optimization was first introduced in the field of game theory by Heinrich Freiherr von Stackelberg with the famous strategic game he described in his book [58] known as the Stackelberg game that consists of two players, a leader and a follower, who compete with each other. The leader makes the first move, and then the follower reacts to the leader's action with his own move. The move of the follower can affect the outcome of both players, therefore, if the leader wants to optimize its objective, then it needs to anticipate the effects of the optimal move of the follower. Hence, one can express this setting as an optimization problem, corresponding to the leader's problem that contains a nested optimization problem corresponding to the follower's optimization problem.

Even though this problem was discussed by Stackelberg in 1934, it was not mathematically formulated until 1973 when the original formulation for bilevel programming appeared in a research paper by Bracken and McGill [59]. The general formulation of the bilevel programming problem is shown in formulation (1.1).

$$
\begin{aligned}
\min_{x_1} \quad & F_1(x_1, x_2) \\
\text{s.t.} \quad & G_1(x_1, x_2) \le 0 \\
& H_1(x_1, x_2) = 0 \\
& x_2 \in \arg\min_{x_2}\left\{F_2(x_1, x_2) : G_2(x_1, x_2) \le 0, H_2(x_1, x_2) = 0\right\} \\
& x_1 \in \mathbb{R}^n, \quad x_2 \in \mathbb{R}^m
\end{aligned}
\tag{1.1}
$$

Since the early 1980s, many algorithms have been proposed for the solution of continuous bilevel problems with many approaches exploiting the Karush–Kuhn–Tucker (KKT) optimality condition of the lower level problem, to transform the bilevel problem into a single level problem. A small indicative list of algorithms for the solution of continuous bilevel problems is presented in Table 1.3, while a small numerical ex-

Table 1.3: Indicative list of solution algorithms for different classes of continuous bilevel optimization problems.

Problem class	Type of algorithm		Reference
LP|LP	Extreme point	Vertex enumeration	Candler and Townsley [60], Bard [61], Tuy et al. [62]
		Kth-best algorithm	Bialas and Karwan [63], Shi et al. [64]
	Branch and Bound	Reformulation	Bard and Falk [65], Fortuny-Amat and McCarl [66]
	Complementary Pivot	SLCP algorithm	Judice and Faustino [67, 68]
	Global Optimisation		Visweswaran et al. [69]
LP|QP	Branch and Bound	Enumeration	Bard and Moore [70]
	Complementary Pivot	SLCP algorithm	Judice and Faustino [71]
QP|QP	Extreme point	Descent approach	Vicente et al. [72]
	Branch and Bound		Bard [73], Al-Khayyal et al. [74], Visweswaran et al. [75], Edmunds and Bard [69]
NLP|NLP	Penalty Function		Aiyoshi and Shimizu [76]
	Grid search		Bard [61, 77]
	Simulated annealing		Sahin and Ciric [78]
	Trust region		Marcotte et al. [79], Colson et al. [80]
	Genetic Algorithm		Yin [81]
	Particle Swarm Optimization		Gao et al. [8]
	KKT or Fritz-John		Dempe and Zemkoho [82, 83], Tsoukalas et al. [84], Dempe and Franke [85], Hansen et al. [86], Mitsos et al. [87]
	Branch-and-bound	Branch-and-Sandwich	Kleniati and Adjiman [88, 89]

ample of a linear continuous bilevel optimization problem solved using the KKT optimality approach is presented in the next section.

1.2.1.1 Numerical example
Consider the following continuous bilevel linear optimization problem:

$$
\begin{aligned}
&\min_{x} \quad 10x + 3y \\
&\text{s.t.} \quad \min_{y} \quad -x - y \\
&\qquad\quad \text{s.t.} \quad -x + 3y - 6 \leq 0 \quad \text{(C1)} \\
&\qquad\qquad\qquad\ -x - 3y - 6 \leq 0 \quad \text{(C2)} \\
&\qquad\qquad\qquad\ \ x + 10y - 10 \leq 0 \quad \text{(C3)} \\
&\qquad\qquad\qquad\ \ x + y - 4 \leq 0 \quad \text{(C4)} \\
&\qquad\qquad\qquad\ \ x - 4y - 8 \leq 0 \quad \text{(C5)}
\end{aligned}
\tag{1.2}
$$

where x is the upper level continuous variable, y is the lower level continuous variable and C1 to C5 are the names of each constraint.

Figure 1.6 is a visualization of the feasible space of the bilevel optimization problem (1.2). As it can be seen in Figure 1.6, and since all constrains are linear, the feasible space of the original optimization problem is a convex polyhedron.

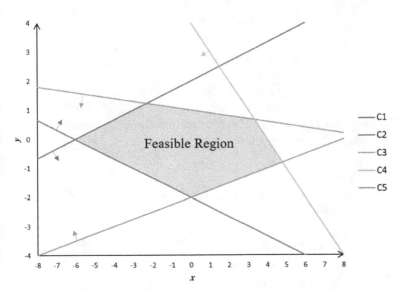

Figure 1.6: Feasible region of bilevel optimization problem (1.2).

The first step in solving this bilevel problem is to derive the KKT optimality conditions of the lower level problem.

Stationarity condition

$$-1 + \begin{bmatrix} u_1 \\ u_2 \\ u_3 \\ u_4 \\ u_5 \end{bmatrix}^T \begin{bmatrix} 3 \\ -3 \\ 10 \\ 1 \\ -4 \end{bmatrix} = 0 \qquad (1.3)$$

Primal feasibility

$$\begin{aligned}
-x + 3y - 6 &\le 0 \\
-x - 3y - 6 &\le 0 \\
x + 10y - 10 &\le 0 \\
x + y - 4 &\le 0 \\
x - 4y - 8 &\le 0
\end{aligned} \qquad (1.4)$$

Dual feasibility

$$u_i \geq 0, \quad i = 1, 2, 3, 4 \tag{1.5}$$

Complementary slackness

$$
\begin{aligned}
u_1(-x + 3y - 6) &= 0 \\
u_2(x - 3y - 6) &= 0 \\
u_3(x + 10y - 10) &= 0 \\
u_4(x + y - 4) &= 0 \\
u_5(x - 4y - 8) &= 0
\end{aligned} \tag{1.6}
$$

Then the KKT conditions derived are substituted into the upper level problem, resulting in formulation (1.7) presented below:

$$
\begin{aligned}
\min_{x,y,u_i} \quad & 10x + 3y \\
\text{s.t.} \quad & -1 + 3u_1 - 3u_2 + 10u_3 + u_4 - 4u_5 = 0 \\
& -x + 3y - 6 \leq 0 \\
& -x - 3y - 6 \leq 0 \\
& x + 10y - 10 \leq 0 \\
& x + y - 4 \leq 0 \\
& x - 4y - 8 \leq 0 \\
& u_i \geq 0, \quad i = 1, 2, 3, 4 \\
& u_1(-x + 3y - 6) = 0 \\
& u_2(x - 3y - 6) = 0 \\
& u_3(x + 10y - 10) = 0 \\
& u_4(x + y - 4) = 0 \\
& u_5(x - 4y - 8) = 0
\end{aligned} \tag{1.7}
$$

The reformulated single level problem (1.7) has nonlinear constraints, from the complementary slackness condition in equations (1.6), and a nonconvex feasible space, illustrated in Figure 1.7.

Therefore, global optimization approaches are needed to solve the reformulated single-level optimization problem to global optimality.

In this simple case, the optimal solution lies at $x^* = -6$ and $y^* = 0$, with the upper level objective being equal to -60 and the lower level objective being equal to 6.

1.2.2 Bilevel mixed-integer optimization

Bilevel decision making problems, such as the case of production and distribution planning introduced in Section 1.1.1, can involve decisions in both discrete and con-

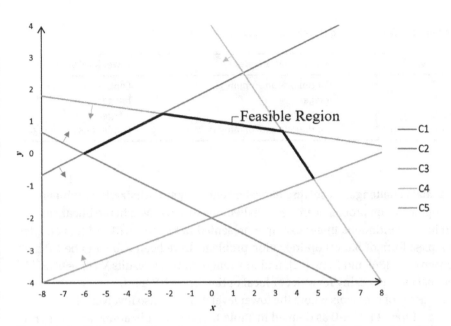

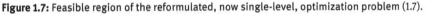

Figure 1.7: Feasible region of the reformulated, now single-level, optimization problem (1.7).

tinuous variables. Problems belonging in this class are referred to as mixed-integer bilevel optimization problems (B-MIP), and have the general form of

$$
\begin{aligned}
&\min_{x_1, y_1} && F_1(x_1, x_2, y_1, y_2) \\
&\text{s. t.} && G_1(x_1, x_2, y_1, y_2) \le 0 \\
& && H_1(x_1, x_2, y_1, y_2) = 0 \\
& && x_2, y_2 \in \arg\min_{x_2, y_2}\big\{F_2(x_1, x_2, y_1, y_2) : G_2(x_1, x_2, y_1, y_2) \le 0, H_2(x_1, x_2, y_1, y_2) = 0\big\} \\
& && x_1 \in \mathbb{R}^m, \quad x_2 \in \mathbb{R}^n, \quad y_1 \in \mathbb{Z}^p, \quad y_2 \in \mathbb{Z}^q
\end{aligned}
\tag{1.8}
$$

where x_1 is a vector of the upper level problem continuous variables, y_1 is a vector of the upper level integer variables, x_2 is a vector of the lower level problem continuous variables and y_2 is a vector of the lower level integer variables.

The general formulation of the mixed-integer bilevel programming problem (1.8), can be divided into a number of different classes of problems, based on where the integer variables appear in the problem. Table 1.4 classifies these problems into four categories that can be expanded to cover the linear, quadratic and nonlinear subclass of each type, as identified by Gumus and Floudas [90] and Avraamidou and Pistikopoulos [91].

Table 1.4: Class Types of mixed-integer bilevel programming problems.

Problem type	Upper level variables	Lower level variables
Type 1-B	Continuous and/or integer	Continuous
Type 2-B	Integer	Integer
Type 3-B	Continuous	Integer
Type 4-B	Continuous and/or integer	Continuous and integer

1.2.2.1 Main challenges in solving bilevel mixed-integer optimization problems

Bilevel optimization problems are very challenging to solve even in the linear case, as shown in the continuous linear example presented in Section 1.2.1.1 of this chapter. The simplest form of bilevel optimization problems have been shown to be NP-hard by Hansen et al. [86] and Deng [92], and to strengthen these results Vicente et al. [72] proved that even checking for strict or local optimality is NP-hard.

For classes of problems where the lower level problem also involves discrete variables (i. e., Types 2-B to 4-B as defined in Table 1.4), the complications are further increased, typically requiring global optimization methods for its solution and often resulting to approximate solutions. The major difficulty for this class of problems arises from the fact that conventional solution methods for continuous bilevel problems are no longer applicable when integer variables exist at the lower level. As discussed in Section 1.2.1, one of the most widely used solution approaches for continuous bilevel problems with convex objective functions and constraints, is the transformation of the problem to a single level problem using the Karush–Kuhn–Tucker (KKT) optimality conditions. Since this method requires gradient information for the lower level problem (see the numerical example in Section 1.2.1.1), it is not directly applicable to bilevel problems with integer variables on the lower level, even though in some cases there is merit in using them, as shown by Gumus and Floudas [90], Saharidis and Ierapetritou [93] and Mitsos [94]. Also, the branch and bound rules used to solve mixed-integer problems cannot be directly or effectively applied to mixed-integer bilevel problems [70].

In the literature, methods developed for the solution of mixed-integer bilevel problems have mainly addressed the linear Type 1-B and 2-B problems. Table 1.5[1] summarizes some of the most important solutions methods for bilevel mixed-integer linear problems of class Types 1-B and 2-B in the open literature, while Table 1.6 summarizes the solution approaches for classes of Type 3-B and 4-B. Table 1.7[2] summarizes approaches for the solution of bilevel mixed-integer nonlinear problems.

---

1 The **Notes** column in the Tables 1.5, 1.6 and 1.7 represents important features, limitations or advantages of the works as written in each individual manuscript.

2 See footnote 1.

Table 1.5: Indicative list of previous work on bilevel mixed-integer **linear** optimization of Type 1-B and 2-B (see definition in Table 1.4).

Type	Algorithm	Reference	Notes
Type 1-B	Branch and Bound	Wen and Yang [97]	Heuristic approach, only integer optimization variables are allowed in the upper level
	Tabu search	Wen and Huang [98]	Only integer optimization variables in the upper level. Approximate solution
	Multiparametric Programming	Faisca et al. [99]	Exact
	Benders decomposition	Caramia and Mari [100]	ϵ-optimal
		Fontaine and Minner [101]	
Type 2-B	Penalty Function	Vicente et al. [102]	Also provided theory for Type 1
	Branch and Bound	Bard and Moore [103]	Implicit Enumeration. Assume all binary and no constraints in the upper level
	Chvatal-Gomory cuts (Cutting plane)	Dempe [104]	Generates a lower bound to the problem
	Branch and Cut (Cutting plane)	DeNegre and Ralphs [105]	Based on Bard and Moore [103]. All binary
	Genetic Algorithm	Nishizaki and Sakawa [106]	Approximate solutions
	Evolutionary Algorithm	Handoko et al. [107]	Global optimality is not guaranteed

1.2.3 Trilevel optimization

Optimization problems involving three decision makers at three different decision levels are referred to as trilevel optimization problems. The first decision maker (upper level; leader) is solving an optimization problem, which includes in its constraint set another optimization problem solved by a second decision maker (second-level follower), and in turn the optimization problem of the second decision maker includes a third optimization problem in its constraint set solved by a third decision maker (third-level follower).

Trilevel decision making problems can involve both discrete and continuous decision variables. Problems in this class are referred to as trilevel mixed-integer optimization problems (T-MIP), and have the general form of (1.9).

Table 1.6: Indicative list of previous work on bilevel mixed-integer **linear** optimization of Type 3-B and 4-B (see definition in Table 1.4).

Type	Algorithm	Reference	Notes
Type 3-B	Branch and Cut	Dempe and Kue [108]	Lower level variables cannot affect the upper level constraints. Also for Type 2 problems
	Polynomial Approximation	Dempe [109], Dempe et al. [110]	Cutting plane, approximate
	Parametric integer programming	Koppe et al. [111]	Cannot guarantee optimality. Theory for Type 2 also provided
Type 4-B	Branch and Bound	Moore and Bard [112]	Implicit Enum. Cannot guarantee optimality
	Penalty Function	Dempe et al. [113]	Approximate local solutions
	Benders decomposition	Saharidis and Ierapetritou [93]	ϵ-optimal. Leader controls all binary variables
	Branch and Bound	Xu and Wang [114], Xu [115]	Only integer optimization variables in the upper level
		Caramia and Mari [100]	
	Lagrangean relaxation	Rahmani and MirHassani [116]	Lower level variables cannot appear in the constraints of the upper level
	Projection-based Reformulation	Yue and You [117] Zeng and An [118]	ϵ-optimal
	Row-and-column generation	Poirion et al. [119]	ϵ-optimal
	Branch-and-Cut	Fischetti et al. [120–122]	Exact. Leader variables that influence the follower decisions are all integer

$$
\begin{aligned}
\min_{x_1, y_1} \quad & F_1(x, y) \\
\text{s.t.} \quad & G_1(x, y) \leq 0 \\
& H_1(x, y) = 0 \\
& \min_{x_2, y_2} \quad F_2(x, y) \\
& \text{s.t.} \quad G_2(x, y) \leq 0 \\
& \qquad H_2(x, y) = 0 \\
& \qquad \min_{x_3, y_3} \quad F_3(x, y) \\
& \qquad \text{s.t.} \quad G_3(x, y) \leq 0 \\
& \qquad \qquad H_3(x, y) = 0 \\
& x = \begin{bmatrix} x_1^T & x_2^T & x_3^T \end{bmatrix}^T, \quad y = \begin{bmatrix} y_1^T & y_2^T & y_3^T \end{bmatrix}^T \\
& x \in \mathbb{R}^n, \quad y \in \mathbb{Z}^m
\end{aligned}
\tag{1.9}
$$

Table 1.7: Indicative list of previous work on bilevel mixed-integer **nonlinear** optimization.

Problem Type	Algorithm	Reference	Notes
Type 1-B	Branch and Bound	Edmunds and Bard [123]	Lower level is convex quadratic
		Gumus and Floudas [95]	Approximate
Type 2-B	Parametric Analysis	Jan and Chern [124]	Only for separable and monotone constraints and objective
	Fuzzy Programming	Emam [125]	Pareto optimal solution
Type 4-B	Simulated Annealing	Sahin and Ciric [78]	Near global solutions
	Branch and Bound	Gumus and Floudas [90]	Cannot solve the general Type 4 B-MINLP problem as the lower level must be linear in continuous variables
	Genetic Algorithms	Hecheng and Yuping [126]	Lower level functions are separable or convex
		Li and Wang [127]	
		Arroyo and Fernandez [128]	Near optimal solutions
	Multiparametric Programming	Dominguez and Pistikopoulos [129]	Reformulation via convex hull Approximate
	Branch and Sandwich	Kleniati and Adjiman [130]	Functions are twice differentiable when the integrality condition is relaxed. ϵ-optimal
	Bounding	Mitsos [94]	ϵ-optimal
	Value-Function-Based	Lozano and Smith [131]	Requires all upper-level variables to be integer
	Gray-box Optimization	Beykal et al. [132]	Requires all upper-level variables to be continuous. No optimality guarantee

where x is a vector of the continuous problem variables and y is a vector of the discrete problem variables. The variable's subscripts indicate the number of the optimization level the decision variables belong to, with 1 corresponding to the first decision level (leader), 2 to the second decision level (first follower) and 3 to the third decision level (second follower).

The general formulation of the T-MIP problem (1.9), corresponds to a number of different subclasses of problems. Similar to the classification for mixed-integer bilevel optimization problems in Section 1.2.2, Table 1.8 classifies the subclasses of trilevel mixed integer optimization problems into three categories depending on the type of variables in each optimization level.

Table 1.8: Class Types of mixed-integer trilevel programming problems.

Type	Level 1 variables	Level 2 and 3 variables
Type 1-T	Integer (and continuous)	Continuous
Type 2-T	Continuous and/or integer	Integer
Type 3-T	Continuous and/or integer	Continuous and integer

1.2.3.1 Main challenges in solving trilevel mixed-integer optimization problems

Multilevel programming problems are very challenging to solve even for the case of two linear continuous decision levels (see discussion in Section 1.2.2.1). The computational complexity of multilevel problems with more than two decision levels was discussed by Blair [133] who noted that the difficulty of solving multi-level linear problems increases significantly when the number of levels increases to more than two.

Solution approaches developed for bilevel programming are not necessarily applicable to trilevel optimization problems. Conventional bilevel solution strategies, such as the substitution of the lower level problem with its Karush–Kuhn–Tucker optimality conditions, typically transform the bilevel problem into a single-level nonlinear (and nonconvex) optimization problem [134] (Section 1.2.1.1). In the case of linear trilevel problems such approaches can be used [135], but for other classes of trilevel problems such approaches will fail to reduce the problem to a tractable single-level problem.

For classes of problems where the lower level problems involve discrete variables, the difficulty further increases, typically requiring global optimization methods and often resulting in approximate solutions [93, 112–114].

Solution approaches presented in the literature for trilevel problems have addressed a very restricted class of problems, mainly linear continuous problems, with only a few attempts to solve problems with integer variables. Table 1.9 summarizes key solution methods for linear problems with three or more decision levels that appear in the open literature. Solution methods for problem class Type 1-T were not found in the literature and, therefore, this class was excluded from Table 1.9. Table 1.10 summarizes key solution methods for continuous nonlinear problems with three or more decision levels that appear in the open literature.

1.2.4 Multifollower optimization

Optimization problems involving a leader with multiple followers are referred to as bilevel multifollower programming problems: the first decision maker (upper level; leader) is solving an optimization problem, which includes in its constraint set other optimization problems solved by second-level decision makers (lower level problems; followers). In recent years, leader-follower games have attracted a growing interest not just in game theory, but also across a broad range of research communities.

Table 1.9: Indicative list of previous work on multilevel **linear** optimization problems with three or more optimization levels.

Type	Algorithm	Reference	Note
Continuous	"Kth-Best" algorithm	Wen and Bialas [136]	For trilevel problems, exact and global
	Cutting plane	Bard [77]	For trilevel problems, exact and global
	Fuzzy programming	Lai [137] Pramanik and Roy [138] Sakawa et al. [139] Shih et al. [140]	For multilevel problems, sub-optimal solutions
	Penalty function	White [141]	For trilevel problems, exact and global
Type 2-T	Tabu search	Sakawa and Matsui [142]	For multilevel problems Suboptimal solutions Only integer variables allowed
	Genetic algorithms	Sakawa et al. [143]	For multilevel problems Suboptimal solutions Only integer variables allowed
Type 3-T	Decomposition algorithm	Yao et al. [42]	For trilevel problems of min-max-min structure, exact and global

Table 1.10: Indicative list of previous work on **continuous nonlinear** multilevel optimization problems with three or more optimization levels.

Class	Algorithm	Reference	Note
Continuous Quatratic	Multiparametric programming	Faisca et al. [144]	Exact and global
Continuous Nonlinear	Particle swarm optimization	Han et al. [145]	Suboptimal solutions
	Evolutionary algorithm	Woldemariam and Kassa [146]	Suboptimal solutions
	Multiparametric programming (B&B)	Kassa and Kassa [147]	Approximate global optimum

Bilevel multifollower decision making problems, such as the problem of planning and scheduling integration introduced in Section 1.1.4, can involve decisions in both discrete and continuous variables. Problems belonging in this class are referred to as mixed-integer bilevel multifollower optimization problems (BMF-MIP), and have the general form of (1.10).

$$\min_{x_1, y_1} \quad F_1(x, y)$$
$$\text{s. t.} \quad G_1(x, y) \le 0$$
$$\min_{x_{2,a}, y_{2,a}} \quad F_{2,a}(x, y)$$
$$\text{s. t.} \quad G_{2,a} \le 0$$
$$\min_{x_{2,b}, y_{2,b}} \quad F_{2,b}(x, y)$$
$$\text{s. t.} \quad G_{2,b} \le 0$$
$$\vdots$$
$$\min_{x_{2,n}, y_{2,n}} \quad F_{2,n}(x, y)$$
$$\text{s. t.} \quad G_{2,n} \le 0$$
$$x = \begin{bmatrix} x_1^T & x_{2,a}^T & x_{2,b}^T & \cdots & x_{2,n}^T \end{bmatrix}^T, \quad x \in \mathbb{R}^n$$
$$y = \begin{bmatrix} y_1^T & y_{2,a}^T & y_{2,b}^T & \cdots & y_{2,n}^T \end{bmatrix}^T, \quad y \in \mathbb{Z}^m$$

$$(1.10)$$

where x_1 is a vector of the upper level continuous problem variables, y_1 is a vector of the upper level integer variables, $x_{2,a}$ to $x_{2,n}$ are vectors of the lower levels continuous problem variables and $y_{2,a}$ to $y_{2,n}$ are vectors of the lower levels integer variables. Note that decision makers $2, a$ to $2, n$ all belong to the same optimization level.

Most research on bilevel optimization mainly addressed the case of a single-follower. The problem of a single leader with multiple followers has not received a lot of attention from the research community with some attempts to solve the linear continuous case [135, 148–151], the nonlinear continuous case [152, 153] and very limited heuristic approaches for the solution of nonlinear mixed-integer multifollower problems [154] that do not guarantee optimality.

1.3 Introduction to multiparametric optimization

Multiparametric optimization is a type of mathematical optimization, where the optimization problem is solved as a function of one or multiple parameters.

In general, multiparametric programming considers the following type of optimization problem:

$$z(\theta) = \underset{x \in \mathbb{R}^n}{\text{minimize}} \quad f(x, \theta) + \tilde{f}(\theta)$$
$$\text{subject to} \quad g(x, \theta) \le 0$$
$$\theta \in \Theta \subset \mathbb{R}^q,$$

$$(1.11)$$

where the parameter space Θ is assumed to be compact.

Note that the function $\tilde{f}(\theta)$ acts as a scaling function, as it can be added to any multiparametric problem formulation without altering the solution $x(\theta)$. Thus, it is omitted from the subsequent formulations for simplicity.

Multiparametric programming is founded on the assumptions and principles of the basic sensitivity theorem as presented in Fiacco (1983). Based on that, the active set in the neighborhood of a nominal parameter vector, θ^*, with a corresponding optimal solution x^* and Lagrange multipliers λ^* of an optimization program, remains unchanged. Consequently, parametric areas (critical regions) with the same active set are created, each characterized with a distinct set of Karush–Kuhn–Tucker (KKT) conditions. Since the optimality conditions remain the same, the resulting system of equations derived from the KKT conditions can be analytically solved, and hence the optimal solution with respect to the varying parameters for the whole parameter space is explicitly constructed.

In the following subsections, we will go over and summarize the developments for certain classes of multiparametric programming problems that will be used for the solution of multilevel optimization problems in the following chapters. More specifically, we will visit the formulation and solution methods of multiparametric linear, quadratic, mixed-integer linear and mixed-integer quadratic optimization problems.

1.3.1 Multiparametric linear programming (mp-LP)

In general, a multiparametric linear programming (mp-LP) problem is defined as

$$
\begin{aligned}
z(\theta) = \underset{x \mathbb{R}^n}{\text{minimize}} \quad & c^T x \\
\text{subject to} \quad & Ax \le b + F\theta \\
& \theta \in \Theta = \left\{ \theta \in \mathbb{R}^q \mid CR_A\,\theta \le CR_b \right\},
\end{aligned}
\tag{1.12}
$$

The solution of problem (1.12) is given by the partitioning of the feasible parameter space $\Theta_f \subseteq \Theta$ into polytopic regions, called critical regions, each of which is associated with the optimal solution $x(\theta)$ and objective function $z(\theta)$, both of which are affine functions of θ (illustrated in Figure 1.8).

Problem (1.12) can be considered the easiest multiparametric programming problem to solve, and it was the first multiparametric programming problem for which a rigorous algorithm appeared in the open literature [155].

1.3.1.1 Degeneracy
The issue of degeneracy in mp-LP problems has intrigued several researchers. The two types of degeneracy typically considered are primal and dual degeneracy (see Figure 1.9) [162]. While primal degeneracy can be handled in a straightforward way by identifying the correct subset of active constraints, which generate a full-dimension parametric solution, the presence of dual degeneracy is more intricate,

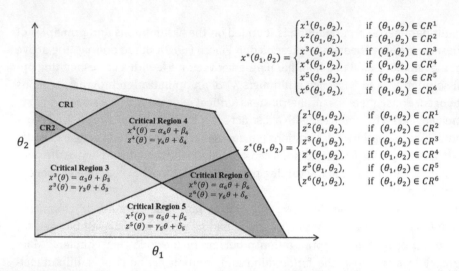

$$x^*(\theta_1, \theta_2) = \begin{cases} x^1(\theta_1, \theta_2), & \text{if } (\theta_1, \theta_2) \in CR^1 \\ x^2(\theta_1, \theta_2), & \text{if } (\theta_1, \theta_2) \in CR^2 \\ x^3(\theta_1, \theta_2), & \text{if } (\theta_1, \theta_2) \in CR^3 \\ x^4(\theta_1, \theta_2), & \text{if } (\theta_1, \theta_2) \in CR^4 \\ x^5(\theta_1, \theta_2), & \text{if } (\theta_1, \theta_2) \in CR^5 \\ x^6(\theta_1, \theta_2), & \text{if } (\theta_1, \theta_2) \in CR^6 \end{cases}$$

$$z^*(\theta_1, \theta_2) = \begin{cases} z^1(\theta_1, \theta_2), & \text{if } (\theta_1, \theta_2) \in CR^1 \\ z^2(\theta_1, \theta_2), & \text{if } (\theta_1, \theta_2) \in CR^2 \\ z^3(\theta_1, \theta_2), & \text{if } (\theta_1, \theta_2) \in CR^3 \\ z^4(\theta_1, \theta_2), & \text{if } (\theta_1, \theta_2) \in CR^4 \\ z^5(\theta_1, \theta_2), & \text{if } (\theta_1, \theta_2) \in CR^5 \\ z^6(\theta_1, \theta_2), & \text{if } (\theta_1, \theta_2) \in CR^6 \end{cases}$$

CR1

CR2

Critical Region 4
$x^4(\theta) = \alpha_4\theta + \beta_4$
$z^4(\theta) = \gamma_4\theta + \delta_4$

Critical Region 3
$x^3(\theta) = \alpha_3\theta + \beta_3$
$z^3(\theta) = \gamma_3\theta + \delta_3$

Critical Region 6
$x^6(\theta) = \alpha_6\theta + \beta_6$
$z^6(\theta) = \gamma_6\theta + \delta_6$

Critical Region 5
$x^5(\theta) = \alpha_5\theta + \beta_5$
$z^5(\theta) = \gamma_5\theta + \delta_5$

Figure 1.8: A schematic representation of the polytopic regions (Critical Regions, CR) created when parametrically solving an optimization problem with two uncertain parameters (θ_1 and θ_2).

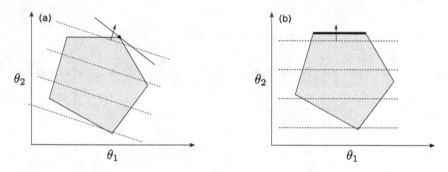

Figure 1.9: A schematic representation of (a) primal and (b) dual degeneracy. In primal degeneracy, the number of active constraints (i. e., inequality constraints where equality holds) is nonunique, while in dual degeneracy the (primal) solution x is nonunique.

as the nonuniqueness of the solution could lead to overlapping critical regions and noncontinuous solutions. The three approaches presented in the literature can be summarized as follows:

Reformulation as mp-QP [163]: When dual degeneracy is identified by the presence of a nonunique primal solution, an equivalent strictly convex mp-QP problem can be formulated.

Graph/Cluster evaluation [155, 164]: In Gal and Nedoma [155], it was shown that the solution to a mp-LP problem is given by a connected graph, where the nodes are the different active sets and the connections are given by the application of a single step of the dual simplex algorithm. Similarly, Olaru and Dumur [164] considers the dual of the mp-LP problem as a parametrized vertex problem, and identifies clusters of connected vertices equivalent to the connections in Gal and Nedoma [155]. When dual degeneracy occurs, multiple disconnected graphs/clusters can occur, only one of which represents the continuous solution of the mp-LP problem across the entire feasible parameter space [164]. Thus, the problem of dual degeneracy is reduced to a graph selection problem.

Lexicographic perturbation [165]: A lexicographic perturbation is based on the concept that a "sufficiently small" perturbation of a degenerate problem will render it nondegenerate [166]. However, this perturbation is not applied but only its effect onto the solution is studied, thus avoiding numerical instabilities.

Handling degeneracy in multiparametric problems can be very useful in obtaining both pessimistic and optimistic solutions of multilevel optimization problems. More details regarding this will be discussed in Chapter 2 and Section 2.1.

1.3.1.2 Critical region definition

In linear programming (LP), the term "basic solution" is a result of the use of the simplex algorithm and identifies the solution as a vertex of the feasible space, which is uniquely defined by the indices of the constraints, which form the vertex. However, with the emergence of interior-point methods, as well as in the face of degeneracy, it cannot be guaranteed that the solution obtained from a LP solver is a basic solution leading to a full-dimensional critical region. As the classical definition of the critical region is directly tied to the active set (i. e., the indices of the constraints which form the vertex), several researchers have considered alternative definitions of critical regions.

The main theme is thereby to identify an appropriate invariancy set over the parametric space. The three sets typically considered are [167]:

Optimal basis invariancy [168]: This invariancy refers to the classical definition of the critical region as a set of active constraints, which form a basic solution. The main issue with this approach occurs in the case of degeneracy (see Section 1.3.1.1), which might lead to lower-dimensional or overlapping regions.

Support set invariancy [169–171]: Given the LP problem formulation

$$\begin{aligned}
\underset{x \in \mathbb{R}^n}{\text{minimize}} \quad & c^T x \\
\text{subject to} \quad & Ax = b \\
& x \geq 0,
\end{aligned} \tag{1.13}$$

the support set is defined as $\sigma(x) = \{i \mid x_i > 0\}$. The concept of support set invariancy describes the region of the parameter space for which the same support set remains optimal. It can be shown that this eliminates the issue of degeneracy, as the support set is independent of the active constraints.

Optimal partition invariancy [170, 172–174]: The optimal partition is given by the cone, which is spanned from the solution found in the directions of the inactive constraints.

1.3.2 Multiparametric quadratic programming (mp-QP)

In general, a multiparametric quadratic programming (mp-QP) problem is defined as

$$
\begin{aligned}
z(\theta) = \underset{x \in \mathbb{R}^n}{\text{minimize}} \quad & (Qx + H\theta + c)^T x \\
\text{subject to} \quad & Ax \le b + F\theta \\
& \theta \in \Theta = \left\{ \theta \in \mathbb{R}^q \mid CR_A\, \theta \le CR_b \right\},
\end{aligned}
\tag{1.14}
$$

where the matrices have appropriate dimensions and Q is symmetric positive definite. The solution to problem (1.14) is given by the partitioning of the feasible parameter space $\Theta_f \subseteq \Theta$ into polytopic regions, called critical regions, each of which is associated with the optimal solution $x(\theta)$ and objective function $z(\theta)$, which are affine and quadratic functions of θ, respectively. Unlike the mp-LP algorithm in Gal and Nedoma [155], which was purely based on simplex-based arguments, the solution of mp-QP problems required the introduction of a new body of theory, namely the basic sensitivity theorem [175].[3] It thereby states that under some mild conditions[4] the change of the optimal solution to a convex optimization can be calculated based on the problem structure and the optimal solution itself. Although the theorem itself appeared in 1976, it was not until the year 2000, when it was shown that for the case of mp-QP problems the optimal solution $x(\theta)$ is an affine function of θ over a given critical region, which is a polytope. Thus, it is conceptually equivalent to mp-LP problems, except for the quadratic nature of the optimal objective function $z(\theta)$ over the critical region.

Therefore, most algorithms which are applicable to mp-QP problems are inherently also applicable to mp-LP problems.

1.3.2.1 Solution procedures

Based on the results from the basic sensitivity theorem, a string of papers in 2000 [176–178], as well as the famous paper by Bemporad et al. from 2002 [179], described a geometrical approach for the solution of mp-QP problems, which relies on the exploration of the parameter space by moving from one critical region to another. As the

---

3 Note that it is also possible to solve mp-QP problems by solving the Karush–Kuhn–Tucker conditions parametrically given a candidate active set.

4 The conditions are: (a) second-order sufficient condition (SOSC), (b) strict complementary slackness (SCS) and (c) linear independence constraint qualification (LICQ). Note that (a) and (b) are inherently fulfilled by strictly convex mp-QP problems.

initial algorithm was shown to generate a large number of artificial cuts [159], new algorithms based on the geometrical principle were described, based on variable step-size [180], inference of the active set of adjacent critical regions [159, 181] and combination of the inference of the active set with the original algorithm [160, 161]. However, in Spjøtvold et al. [161], Spjotvold et al. [160] a fundamental limitation of the most efficient geometrical algorithms was uncovered, namely the facet-to-facet property. It states that the geometrical algorithms [159, 180] are only guaranteed to yield the correct solution if the linear independent constraint qualification (LICQ) is fulfilled on the facet of each critical region.[5] As this condition does not hold unconditionally for mp-QP problems, geometrical algorithms which do not produce artificial cuts are generally not guaranteed to yield the complete solution.

Shortly after the appearance of the first papers for the solution of mp-QP problems in 2000, several papers considered a different strategy to the geometrical considerations [182–185]. In particular, it was realized that the solution of a mp-QP problem is uniquely defined by its corresponding active set. Thus, it was suggested to exhaustively enumerate all possible combinations of active sets, an approach that Mayne and Raković termed "reverse transformation" [183]. However, for any but the smallest problems this approach is computationally intractable. This limitation was tackled in 2011, when Gupta et al. showed that if a candidate active set k is infeasible, so is its power set [186]. This led to the development of a branch-and-bound algorithm for the active set combinations, which is not based on the facet-to-facet property and is thus guaranteed to yield the complete solution. Since then, several publications have appeared utilizing symmetry arguments to increase the fathoming efficiency or applying the concept to multiparametric linear complementarity problems [187–190].

Despite the fact that both the geometrical and the combinatorial approach solve the same problem, they explore unrelated aspects of the problem structure. Thus, in an effort to establish a connection between these two approaches, recently the authors generalized the connected-graph approach from Gal and Nedoma [155] to the mp-QP case, as it was shown that the solution to a strictly convex mp-QP problem is given by a connected graph [191]. The nodes are thereby the optimal active sets over the parameter space, and the connections are given by the inference of the adjacent active set [159, 161] or a single step of the dual simplex algorithm [155].

In parallel to these developments, the possibility to reformulate both mp-LP and mp-QP problems into multiparametric linear complementarity problems (mp-LCP) [192] was considered, for which geometrical and combinatorial approaches have been presented [190, 193].

---

5 The theorem also requires the solution not to be dual degenerate, which is always satisfied for mp-QP problems with a positive definite Q.

Several papers have conceptualized other solution strategies based, e. g., graphical derivatives [194] or (parametrized) vertex approaches [195, 196].

1.3.3 Multiparametric mixed-integer linear programming

In general, a multiparametric mixed-integer linear programming (mp-MILP) problem is defined as

$$
\begin{aligned}
z(\theta) = \underset{x,y}{\text{minimize}} \quad & c^T \omega \\
\text{subject to} \quad & Ax + Ey \leq b + F\theta \\
& x \in \mathbb{R}^n, \quad y \in \{0,1\}^p, \quad \omega = \begin{bmatrix} x^T & y^T \end{bmatrix}^T \\
& \theta \in \Theta = \{\theta \in \mathbb{R}^q \mid CR_A\, \theta \leq CR_b\},
\end{aligned}
\tag{1.15}
$$

where the matrices have appropriate dimensions.[6] The solution to problem (1.15) is given by the partitioning of the feasible parameter space $\Theta_f \subseteq \Theta$, which might be nonconvex, into polytopic regions, called critical regions, each of which is associated with the optimal solution $x(\theta)$ and objective function $z(\theta)$, which are affine functions of θ. These properties are a direct result of the fact that the solution of a mp-MILP problem can be viewed as the combination of the solution of several mp-LP problems.

Thus, in order to solve mp-MILP problems, it is sufficient to exhaustively enumerate all possible combinations of binary variables, solve the resulting mp-LP problems and combine the solutions by comparing them to each other [200–202]. However, it is clear that such an approach is computationally intractable for larger numbers of binary variables. Thus, several researchers have investigated techniques to reduce the number of combinations of binary variables considered. The two research directions that have been pursued are thereby a branch-and-bound strategy and a decomposition-based algorithm:

Brand-and-bound strategy: The key idea of this approach is to extend the well-known concepts of branch-and-bound used for the solution of regular MILP problems to the multiparametric case [203, 204].[7] At the root node the binary variables y are relaxed from $y \in \{0,1\}^p$ to $\bar{y} \in [0,1]^p$, i. e., into constrained continuous variables. The resulting mp-LP problem is solved using the approaches in Section 1.3.1, and the child nodes are created where one of the variables in $\bar{y}$ is

---

6 Some authors have also denoted a mp-P featuring the objective function $(H\theta + c)^T \omega$ as a mp-MILP [197, 198]. However, since such an objective function may lead to quadratically constrained critical regions [199], they are fundamentally more complex than problem (1.15), and in fact conceptually equivalent to mp-MIQP problems, which are discussed in the next section.

7 Interestingly, this was the first mp-MILP algorithm presented in the literature in 1997, even before the use of exhaustive enumeration was considered in the year 2000.

fixed to 0 and 1, and the mp-LP problem is solved again. If the solution of a node is shown to be neither feasible nor optimal over the entire parameter space, the node and all its child nodes are discarded.

Decomposition-based strategy: Instead of solving the mp-MILP problem directly in a multiparametric fashion, the decomposition based algorithm relies on the identification of a candidate combination of binary variables *via* the solution of a MILP problem [205, 206]. This candidate is then fixed and the resulting mp-LP problem is solved, and compared with the current best upper bound. At the next iteration, in each critical region a new MILP problem is solved, which searches for a candidate combination of binary variables that yields a better solution in at least one point over the critical region considered. If such a point cannot be found, then the considered critical region is optimal for the original mp-MILP problem.

1.3.3.1 Multiparametric integer linear programming

In the case where $n = 0$ in problem (1.15), i. e., without any continuous variables in the problem formulation, problem (1.15) yields a multiparametric integer linear programming (mp-ILP) problem. As it is a subclass of mp-MILP problems, the same algorithms can be used to solve mp-ILP algorithms. However, in a series of papers, Crema [207–212] developed a generalized approach for the solution mp-ILP algorithms, which utilizes some specific characteristics inherent to mp-ILP problems.

In essence, the lack of continuous variables implies that the objective function stays constant once a candidate combination of binary variables has been found. Thus, there is no need to perform a comparison procedure, and the only way another combination of binary variables might be optimal is when the initial combination becomes infeasible. Thus, it is sufficient to explore the parameter space in a geometric fashion in order to identify feasible combinations of binary variables, which are outside the current feasible space of the solution. However, as Θ_f is possibly nonconvex, the consideration of the entire parameter space needs to be guaranteed, which can be achieved by formulating a single large-scale MILP problem.

1.3.4 Multiparametric mixed-integer quadratic programming

In general, a multiparametric mixed-integer quadratic programming (mp-MIQP) problem is defined as

$$
\begin{aligned}
z(\theta) = \underset{x,y}{\text{minimize}} \quad & (Q\omega + H\theta + c)^T \omega \\
\text{subject to} \quad & Ax + Ey \le b + F\theta \\
& x \in \mathbb{R}^n, \quad y \in \{0,1\}^p, \quad \omega = \begin{bmatrix} x^T & y^T \end{bmatrix}^T \\
& \theta \in \Theta = \{\theta \in \mathbb{R}^q \mid CR_A\, \theta \le CR_b\},
\end{aligned}
\tag{1.16}
$$

where the matrices have appropriate dimensions and Q is symmetric positive definite. The solution to problem (1.16) is given by the partitioning of the feasible parameter space $\Theta_f \subseteq \Theta$, which might be nonconvex, into quadratically constrained regions, called critical regions, each of which is associated with the optimal solution $x(\theta)$ and objective function $z(\theta)$, which are affine and quadratic functions of θ, respectively. The key difference to mp-MILP algorithms is thereby the fact that the critical regions are not polytopic, but quadratically constrained. This is a direct result of the fact the solution of mp-MIQP problems can be viewed as the combination of solution of several mp-QP problems. This combination requires the comparison of the different solutions, which is driven by the quadratic objective function. Thus, the resulting transitions between solutions is potentially quadratic, resulting in quadratically constrained regions.

1.3.4.1 The envelope of solutions

The polytopic nature of the critical regions has many computational advantages, as the software tools for tackling such systems are very efficient. Thus, in order to preserve this polytopic nature, the notion of envelopes of solutions was introduced, where more than one solution is associated with a critical region, thus avoiding the comparison procedure and the generation of quadratically constrained critical regions. In the first paper, which presented a solution procedure for mp-MIQP problems in 2002 [213], the authors did not perform any comparison and simply kept all candidate solutions resulting from the application of the decomposition-based approach (see Section 1.3.3). This remained the state-of-the-art procedure, until Axehill et al. [214, 215] presented an approach, which coupled the branch-and-bound approach with the idea of comparing the solutions over the entire critical region. While this does require the solution of potentially nonconvex QP problems, it significantly reduced the number of solutions stored within each region [199]. Independently, Oberdieck et al. [198] tackled the same problem by using McCormick relaxations to linearize the quadratic parts of the objective functions for the comparison procedure. While this approach significantly increases the number of partitions as a result of the linearization, it also reduced the number of solutions stored within each region Oberdieck and Pistikopoulos [199]. Despite these developments, all of these approaches still generate an envelope of solutions. Only in 2015 the first exact algorithm for the solution of mp-MIQP problems appeared [199], which enabled the handling of quadratically constrained critical regions by applying McCormick relaxations to the critical regions instead of the objective functions to be compared.

Another way to eliminate the presence of envelopes of solutions is the lifting of the solution into a higher-dimensional space, which transforms the solution to a mp-MIQP problem into the form of a mp-QP problem by treating the quadratic terms as extra dimensions [216].

1.3.4.2 Further reading on multiparametric programming

Table 1.11 contains publications of algorithms that can solve the classes of multiparametric problems discussed in this section.

Table 1.11: Multiparametric programming algorithms for the solution of mpLPs, mpMPs, mpMILPs and mpMIQPs.

	mpLP	mpMILP	mpQP	mpMIQP
Gal and Nedoma [155]	X			
Gal [217]	X			
Yuf and Zeleny [218]	X			
Schechter [219]	X			
Acevedo and Pistikopoulos [203]	X	X		
Pertsinidis et al. [220]	X	X		
Pistikopoulos et al. [178]	X		X	
Dua and Pistikopoulos [205]	X	X		
Bemporad et al. [179]	X		X	
Dua et al. [213]	X	X	X	X
Bemporad et al. [157]	X			
Baotic [221]	X		X	
Tøndel et al. [181]	X		X	
Filippi [222]	X			
Jia and Ierapetritou [206]	X	X		
Spjøtvold et al. [161]	X		X	
Jones et al. [166]	X			
Li and Ierapetritou [223]	X	X		
Jones et al. [165]	X			
Faísca et al. [224]	X	X		
Mitsos and Barton [225]	X	X		
Patrinos and Sarimveis [194]	X		X	
Li and Ierapetritou [226]	X	X	X	X
Gupta et al. [186]	X		X	
Feller and Johansen [188]	X		X	
Wittmann-Hohlbein and Pistikopoulos [227]	X	X		
Oberdieck et al. [228]	X	X		
Wittmann-Hohlbein and Pistikopoulos [229]	X	X		
Axehill et al. [215]	X	X	X	X
Bemporad [230]	X		X	
Herceg et al. [190]	X	X	X	X
Oberdieck and Pistikopoulos [231]	X	X	X	X
Oberdieck et al. [191]	X		X	
Charitopoulos et al. [232]	X		X	
Ahmadi-Moshkenani et al. [233]	X		X	
Akbari and Barton [234]	X			
Burnak et al. [235]	X	X	X	

For further reading on multiparametric programming, the reader can also refer to the following books:

1. Pistikopoulos, E. N.; Diangelakis, N. A.; Oberdieck, R. Multi-parametric Optimization and Control; John Wiley & Sons, 2020.
2. Pistikopoulos, E. N., Georgiadis, M. C., Dua, V., Eds. Multi-Parametric Programming; Process Systems Engineering 1; Wiley-VCH: Weinheim, 2011.
3. Pistikopoulos, E. N., Georgiadis, M. C., Dua, V., Eds. Multi-Parametric Model-Based Control; Process Systems Engineering 2; Wiley-VCH: Weinheim, 2011.

1.3.5 Solution of continuous bilevel problems through multiparametric programming

The main idea behind the development of multiparametric algorithms for the solution of bilevel programing problems came out through the following observation:

In a bilevel optimization setting, the lower level optimization problem is parametric in terms of the upper level variables.

This observation gave rise to several methodologies and algorithms for the solution of bilevel problems with the key idea to solve the lower level problem parametrically in terms of the upper level variables [236]. A list of different classes of multilevel problems solved using multiparametric programming is presented in Table 1.12.

Table 1.12: Classes of problems solved using multiparametric programming and corresponding contributions.

Problem class	Reference
LP|LP	Faisca et al. [99]
	Pistikopoulos et al. [237, 238]
LP|QP	Faisca et al. [99]
	Pistikopoulos et al. [237, 238]
QP|LP	Pistikopoulos et al. [237]
QP|QP	Faisca et al. [99]
	Pistikopoulos et al. [237, 238]
LP|LP with uncertainty	Ryu et al. [14]
QP|QP with uncertainty	Faisca et al. [99]
	Pistikopoulos et al. [238]
LP|LP|LP	Faisca et al. [144]
QP|QP|QP	Faisca et al. [144, 239]
	Pistikopoulos et al. [238]
QP|QP,QP	Faisca et al. [144, 239]
	Pistikopoulos et al. [238]
LP|LP,LP,LP	Faisca et al. [144]

The main steps of the algorithms for the solution of all the classes mentioned in Table 1.12 are:

1. Recast the lower level problem as a multiparametric optimization problem, considering the upper level variables as parameters.
2. Solve the resulting multiparametric problem using a suitable solution algorithm (either mpLP or mpQP) to get k critical regions.
3. Substitute each of the k solutions into the leader's problem to formulate k single level optimization problems.
4. Solve the k single level optimization problems and compare the optimal points to select the best one.

1.3.5.1 Further reading on continuous multilevel optimization problems through multiparametric programming

For numerical examples and further reading on the solution approaches for continuous multilevel optimization problems through multiparametric programming, the reader can refer to the papers in Table 1.12, and to Chapter 6, "Bilevel and Multilevel Programming" of the book, Pistikopoulos, E. N., Georgiadis, M. C., Dua, V., Eds. Multi-Parametric Programming; Process Systems Engineering 1; Wiley-VCH: Weinheim, 2011.

1.4 Organization of the book

The remainder of this book is organized in two parts. The first part includes Chapters 2 and 3, while the second part includes Chapters 4 and 5. Part I focuses on the introduction of algorithms for the solution of different classes of multilevel mixed-integer optimization problems, while Part II focuses on application case studies and the computational implementation of the introduced algorithms.

Therefore, the next chapter (Chapter 2) focuses on bilevel mixed-integer optimization. An algorithm based on multiparametric programming is first introduced to solve bilevel mixed-integer linear optimization problems and then expanded for the solution of bilevel mixed-integer quadratic optimization problems and bilevel problems with right-hand side uncertainty. All three algorithms are explained step-by-step through multiple numerical examples.

2 Bilevel mixed-integer optimization

In this chapter, we present global optimization algorithms for the exact and global solution of two classes of bilevel programming problems, namely (i) bilevel mixed-integer linear programming problems (B-MILP) and (ii) bilevel mixed-integer convex quadratic programming problems (B-MIQP), belonging to problem class Type 4 (i. e., containing both integer and continuous variables at both optimization levels), while also applicable to problem classes of Types 1–3 (see Table 1.4 for the definition of bilevel mixed-integer problem classes).

2.1 Mixed-integer linear bilevel optimization

Expanding on the earlier work of Faisca et al. [99, 239], which address continuous bilevel programming problems, the approach that we will focus on here is based upon multiparametric programming theory (see Section 1.3), the Multiparametric Mixed-integer Linear Programming (mp-MILP) algorithm of Oberdieck et al. [198], and the theory for binary parameters in multiparametric programming problems by Oberdieck et al. [240]. The solution method will be first introduced through the general form of the B-MILP problem (2.1), and then illustrated through 3 numerical examples in Section 2.1.1.

$$
\begin{aligned}
\min_{x_1, y_1} \quad & F_1(x, y) = c_1^T x + d_1^T y \\
\text{s.t.} \quad & A_1 x + B_1 y \leq b_1 \\
& x_2, y_2 \in \arg\min_{x_2, y_2}\left\{F_2(x, y) = c_2^T x + d_2^T y : A_2 x + B_2 y \leq b_2\right\} \\
& x = \begin{bmatrix} x_1^T & x_2^T \end{bmatrix}^T, \quad y = \begin{bmatrix} y_1^T & y_2^T \end{bmatrix}^T \\
& x \in \mathbb{R}^n, \quad y \in \mathbb{Z}^m
\end{aligned}
\tag{2.1}
$$

where c_1, d_1, A_1, B_1, b_1 are constant coefficient matrices in the upper level (leader) problem, and c_2, d_2, A_2, B_2, b_2 are constant coefficient matrices in the lower level (follower) problem. It is assumed that upper level optimization variables that appear in the lower level problem, and lower level integer variables are bounded.

As a first step, we establish bounds for all integer and continuous variables, by solving problems (2.2) to (2.5) for upper level variables x_1^α and y_1^β, for all $\alpha \in \{1, \ldots, n_1\}$ and $\beta \in \{1, \ldots, n_2\}$, and problems (2.6) to (2.9) for lower level variables x_1^γ and y_2^δ, for all $\gamma \in \{1, \ldots, n_3\}$ and $\delta \in \{1, \ldots, n_4\}$, in order to obtain bounds on both x, $x^L \leq x \leq x^U$, and y, $y^L \leq y \leq y^U$:

$$
\begin{aligned}
x_1^{\alpha L} = \min \quad & x_1^\alpha \\
\text{s.t.} \quad & A_1 x + B_1 y \leq b_1 \\
& A_2 x + B_2 y \leq b_2
\end{aligned}
\tag{2.2}
$$

https://doi.org/10.1515/9783110760316-002

$$
\begin{aligned}
x_1^{\alpha U} = \min \quad & -x_1^{\alpha} \\
\text{s. t.} \quad & A_1 x + B_1 y \le b_1 \\
& A_2 x + B_2 y \le b_2
\end{aligned}
\tag{2.3}
$$

$$
\begin{aligned}
y_1^{\beta L} = \min \quad & y_1^{\beta} \\
\text{s. t.} \quad & A_1 x + B_1 y \le b_1 \\
& A_2 x + B_2 y \le b_2
\end{aligned}
\tag{2.4}
$$

$$
\begin{aligned}
y_1^{\beta U} = \min \quad & -y_1^{\beta} \\
\text{s. t.} \quad & A_1 x + B_1 y \le b_1 \\
& A_2 x + B_2 y \le b_2
\end{aligned}
\tag{2.5}
$$

$$
\begin{aligned}
x_2^{\alpha L} = \min \quad & x_2^{\alpha} \\
\text{s. t.} \quad & A_1 x + B_1 y \le b_1 \\
& A_2 x + B_2 y \le b_2
\end{aligned}
\tag{2.6}
$$

$$
\begin{aligned}
x_2^{\alpha U} = \min \quad & -x_2^{\alpha} \\
\text{s. t.} \quad & A_1 x + B_1 y \le b_1 \\
& A_2 x + B_2 y \le b_2
\end{aligned}
\tag{2.7}
$$

$$
\begin{aligned}
y_2^{\beta L} = \min \quad & y_2^{\beta} \\
\text{s. t.} \quad & A_1 x + B_1 y \le b_1 \\
& A_2 x + B_2 y \le b_2
\end{aligned}
\tag{2.8}
$$

$$
\begin{aligned}
y_2^{\beta U} = \min \quad & -y_2^{\beta} \\
\text{s. t.} \quad & A_1 x + B_1 y \le b_1 \\
& A_2 x + B_2 y \le b_2
\end{aligned}
\tag{2.9}
$$

Then the B-MILP is transformed into a binary B-MILP by expressing integer variables, $y_1^1 \ldots y_1^{n_2}$ and $y_2^1 \ldots y_2^{n_4}$, in terms of binary 0-1 variables, $y_{1,1}^{\acute{\beta}}, \ldots, y_{1,n_5}^{\acute{\beta}} \in \{0,1\}$ for all β and $y_{2,1}^{\acute{\delta}}, \ldots, y_{2,n_6}^{\acute{\delta}} \in \{0,1\}$ for all δ, by following the formulas presented in Floudas [241], in Section 6.2.1, Remark 1. The acute accent will be omitted in the following steps for simplicity.

As a next step, the lower level problem of the binary B-MILP, is transformed into a mp-MILP problem (2.10), in which the optimization variables of the upper level problem, x_1 and y_2 that appear in the lower level problem, are considered as parameters for the lower level:

$$
\begin{aligned}
\min_{x_2, y_2} \quad & d_2^T y + c_2^T x \\
\text{s. t.} \quad & B_2 y \le b_2 - A_2 x \\
& x^L \le x \le x^U
\end{aligned}
\tag{2.10}
$$

The multiparametric problem (2.10) is solved using the multiparametric solution algorithm presented in Chapter 1, Section 1.3, through the POP® toolbox [242] to provide the complete profile of optimal solutions of the lower level problem as explicit functions of the variables of the higher level problem with corresponding expressions (2.11). POP® toolbox features a state-of-the-art multiparametric programming solver for continuous and mixed-integer linear and quadratic problems. The toolbox is freely available for download at *parametric.tamu.edu* website. More information on POP can be found in Appendix B.

$$
x_2, y_2 = \begin{cases}
\xi_1 = p_1 + q_1 x_1 + r_1 y_1, & \psi_1 \quad \text{if } H_1[x_1^T \ y_1^T]^T \le h_1 \\
\xi_2 = p_2 + q_2 x_1 + r_2 y_1, & \psi_2 \quad \text{if } H_2[x_1^T \ y_1^T]^T \le h_2 \\
\vdots & \qquad \vdots \\
\xi_k = p_k + q_k x_1 + r_k y_1, & \psi_k \quad \text{if } H_k[x_1^T \ y_1^T]^T \le h_k
\end{cases}
\tag{2.11}
$$

where ξ_i are vectors of the lower level (follower) continuous variables and ψ_i are vectors of the lower level integer variables, $H_k[x_1^T \ y_1^T]^T \le h_k$ is referred to as the critical region, CR^k, and k denotes the number of computed critical regions.

The computed solutions (2.11) are then substituted into the upper level problem, which can be solved as a set of single-level deterministic mixed-integer programming problems, (2.12). More specifically, the functions ξ expressing the lower level variables (x_2) in terms of the upper level variables $(x_1$ and $y_1)$, are substituted in the place of lower level variables $(x_2$ and $y_2)$ in the upper level problem, eliminating in this way the lower level variables form the upper level problem. Moreover, the critical region definitions are added to the corresponding single level problems as an additional set of constraints:

$$
\begin{aligned}
z_1 = \min_{x_1, y_1} \quad & c_1^T[x_1^T \ \ \xi_1(x_1, y_1)^T]^T + d_1^T[y_1^T \ \ \psi_1^T]^T \\
\text{s.t.} \quad & A_1[x_1^T \ \ \xi_1(x_1, y_1)^T]^T + B_1[y_1^T \ \ \psi_1^T]^T \le b_1 \\
& H_1[x_1^T \ y_1^T]^T \le h_1 \\
z_2 = \min_{x_1, y_1} \quad & c_1^T[x_1^T \ \ \xi_2(x_1, y_1)^T]^T + d_1^T[y_1^T \ \ \psi_2^T]^T \\
\text{s.t.} \quad & A_1[x_1^T \ \ \xi_2(x_1, y_1)^T]^T + B_1[y_1^T \ \ \psi_2^T]^T \le b_1 \\
& H_2[x_1^T \ y_1^T]^T \le h_2 \\
& \qquad \vdots \\
z_k = \min_{x_1, y_1} \quad & c_1^T[x_1^T \ \ \xi_k(x_1, y_1)^T]^T + d_1^T[y_1^T \ \ \psi_k^T]^T \\
\text{s.t.} \quad & A_1[x_1^T \ \ \xi_k(x_1, y_1)^T]^T + B_1[y_1^T \ \ \psi_k^T]^T \le b_1 \\
& H_k[x_1^T \ y_1^T]^T \le h_k
\end{aligned}
\tag{2.12}
$$

The single-level, deterministic programming problems (2.12) are independent of each other, making it possible to use parallel programming to solve them simultaneously.

The solutions of the above single level MILP problems correspond to different local optimal solutions of the original B-MILP. The final step of the algorithm is to compare all the local solutions to obtain the minimum z that would correspond to the exact and global optimum, z^*, of the original bilevel problem.

The proposed algorithm is summarized in Table 2.1.

Table 2.1: Multiparametric based algorithm for the solution of B-MILP problems.

Step 1	Establish integer and continuous variable bounds
Step 2	Transform the B-MILP into a binary B-MILP
Step 3	Recast the lower level as a mp-MILP, in which the optimization variables of the upper level problem are considered as parameters
Step 4	Solve the resulting mp-MILP problems to obtain the optimal solution of the lower level problem as explicit functions of the upper level variables
Step 5	Substitute each multiparametric solution into the upper level problem to formulate k single level MILP problems
Step 6	Solve all k single level problems and compare their solutions to select the exact and global optimum

Remark (Pessimistic and optimistic solutions). The choice of a pessimistic versus an optimistic solution emerges when the optimal solution of the lower level problem is not unique for the set of optimal upper level variables.

If this degeneracy results because of the lower level binary variables, the solution method described above is able to capture all degenerate solutions and, therefore, supply the decision maker with both the pessimistic and optimistic solutions, and all other solutions.

For the cases where a degeneracy results because of the lower level continuous variables, the multiparametric solution via POP® toolbox is not able to supply the decision maker with the full range of degenerate solution. Even though there are techniques to handle degeneracy in multiparametric problems [155, 163, 164, 166] (see also Section 1.3.1.1), those are not yet implemented in the approach described above.

Therefore, it is assumed that there is a unique optimal solution for the continuous lower level variables corresponding to the upper level optimal solution.

2.1.1 Numerical examples

Three B-MILP numerical examples will be solved to illustrate the use of the proposed algorithm.

2.1.1.1 Example 1: LP-ILP

Consider the following Type 3^1 example taken from Dempe [109]:

$$\min_{y} \quad -x_1 - 2x_2 + 3y_1 + 3.2y_2$$
$$\text{s. t.} \quad -y_1 - y_2 \leq 2$$
$$y_1 + y_2 \leq 2$$
$$-2 \leq y_{1,2} \leq 2$$
$$\min_{x} \quad -x_1 y_1 - x_2 y_2 \qquad (2.13)$$
$$\text{s. t.} \quad -x_1 + 3x_2 \leq 3$$
$$x_1 - x_2 \leq 1$$
$$-x_1 - x_2 \leq -2$$
$$y \in \mathbb{R}^n, \quad x \in \mathbb{Z}^{+m}$$

Step 1: Bounds are established for the unbounded integer variables x_1 and x_2 by solving problems (2.14) to (2.17), resulting to $1 \leq x_1 \leq 3$ and $1 \leq x_2 \leq 2$:

$$x_1^L = \min_{x_1, x_2} \quad x_1$$
$$\text{s. t.} \quad -x_1 + 3x_2 \leq 3$$
$$x_1 - x_2 \leq 1 \qquad (2.14)$$
$$-x_1 - x_2 \leq -2$$

$$-x_1^U = \min_{x_1, x_2} \quad -x_1$$
$$\text{s. t.} \quad -x_1 + 3x_2 \leq 3$$
$$x_1 - x_2 \leq 1 \qquad (2.15)$$
$$-x_1 - x_2 \leq -2$$

$$x_2^L = \min_{x_1, x_2} \quad x_2$$
$$\text{s. t.} \quad -x_1 + 3x_2 \leq 3$$
$$x_1 - x_2 \leq 1 \qquad (2.16)$$
$$-x_1 - x_2 \leq -2$$

$$-x_2^U = \min_{x_1, x_2} \quad -x_2$$
$$\text{s. t.} \quad -x_1 + 3x_2 \leq 3$$
$$x_1 - x_2 \leq 1 \qquad (2.17)$$
$$-x_1 - x_2 \leq -2$$

---

1 See Table 1.4 for the definition of bilevel mixed-integer problem class types.

Step 2: The problem is transformed into a 0-1 binary B-MILP. Following [241], the integer variables x_1 and x_2 can be expressed through binary variables as $x_1 = 1 + x_{1a} + 2x_{1b}$ and $x_2 = 1 + x_{2a}$. Therefore, formulation (2.13) can be reformulated as (2.18):

$$
\begin{aligned}
\min_{y} \quad & -x_{1a} - 2x_{1b} - 2x_{2a} + 3y_1 + 3.2y_2 - 3 \\
\text{s.t.} \quad & -y_1 - y_2 \leq 2 \\
& y_1 + y_2 \leq 2 \\
& -2 \leq y_{1,2} \leq 2 \\
& (x_{1a}, x_{1b}, x_{2a}) \in \arg\Big\{ \min_{x_{1a}, x_{1b}, x_{2a}} \quad (x_{1a} + 2x_{1b} + 1)y_1 + (x_{2a} + 1)y_2 \\
& \qquad\qquad \text{s.t.} \quad -x_{1a} - 2x_{1b} + 3x_{2a} \leq 1 \\
& \qquad\qquad\qquad\quad x_{1a} + 2x_{1b} - x_{2a} \leq 1 \\
& \qquad\qquad\qquad\; -x_{1a} - 2x_{1b} - x_{2a} \leq 0 \Big\} \\
& y \in \mathbb{R}^2, \quad x_{1a}, x_{1b}, x_{2a} \in \{0, 1\}^3
\end{aligned}
\tag{2.18}
$$

Step 3: Then the lower level problem is reformulated as a multiparametric integer linear programming problem (2.19), in which the optimization variables of the upper level problem y_1 and y_2 are considered as parameters:

$$
\begin{aligned}
\min_{x_{1a}, x_{1b}, x_{2a}} \quad & \left(\begin{bmatrix} 1 & 0 \\ 2 & 0 \\ 0 & 1 \end{bmatrix} y \right)^T \begin{bmatrix} x_{1a} \\ x_{1b} \\ x_{2a} \end{bmatrix} + \begin{bmatrix} 1 & 1 \end{bmatrix} y \\
\text{s.t.} \quad & \begin{bmatrix} -1 & -2 & 3 \\ 1 & 2 & -1 \\ -1 & -2 & -1 \end{bmatrix} \begin{bmatrix} x_{1a} \\ x_{1b} \\ x_{2a} \end{bmatrix} \leq \begin{bmatrix} 1 \\ 1 \\ 0 \end{bmatrix} \\
& -2 \leq y_{1,2} \leq 2 \\
& x_{1a}, x_{1b}, x_{2a} \in \{0, 1\}^3
\end{aligned}
\tag{2.19}
$$

Step 4: The multiparametric optimization problem (2.19) is then solved using the appropriate algorithm through POP® toolbox [242] and yields the optimal parametric solution given in Table 2.2. In this example, the parametric solution consists of only one critical region.

Table 2.2: Lower level problem solution of Example 1.

Critical Region	Definition	Objective function	Variable value
CR1	$-2 \leq y_{1,2} \leq 2$	$y_1 + y_2$	$x_{1a} = 0, x_{1ab} = 0, x_{2a} = 0$

Step 5: The solution obtained is then substituted into the upper level problem to formulate one new single-level deterministic linear programming (LP) problem (2.20).

$$\min_{y} \quad -3 + 3y_1 + 3.2y_2$$
$$\text{s. t.} \quad -y_1 - y_2 \leq 2$$
$$y_1 + y_2 \leq 2 \tag{2.20}$$
$$-2 \leq y_{1,2} \leq 2$$
$$y \in \mathbb{R}^2$$

Step 6: Problem (2.20) is solved using CPLEX linear programming solver, and results to the solution presented in Table 2.3. Since only one solution is derived no comparison procedure in this step is needed and the solution listed in Table 2.3 is the exact and global optimal solution of Example 1 (2.13).

Table 2.3: Solution of the single level problem formulated in Example 1.

Objective function	Continues variables	Discrete variables
−9.4	$y_1 = 0, y_2 = -2$	$x_1 = 1, x_2 = 1$

2.1.1.2 Example 2: ILP-ILP

Consider the following Type 2^2 class example taken from [112]:

$$\min_{x} \quad -3 + 3y_1 + 3.2y_2$$
$$\text{s. t.} \quad y \in \arg\{\min_{y} \quad y$$
$$\text{s. t.} \quad x + 2y \leq 10$$
$$x + 2y \leq 10 \tag{2.21}$$
$$2x - y \leq 15$$
$$-2x - 10y \leq -15\}$$
$$x, y \in \mathbb{Z}^{+2}$$

Steps 1 and 2: Bounds are established for all the variables, resulting to $1 \leq x \leq 8$ and $1 \leq y \leq 4$. The problem is then transformed into a 0-1 binary B-ILP problem (2.22), by expressing the integer variables x and y through the binary variables x_1, x_2, x_3, y_1

---

2 See Table 1.4 for the definition of bilevel mixed-integer problem class types.

and y_2 as $x = 1 + x_1 + 2x_2 + 4x_3$ and $y = 1 + y_1 + 2y_2$ [241]:

$$\min_{x_1, x_2, x_3} \quad -x_1 - 2x_2 - 4x_3 - 10y_1 - 20y_2 - 11$$

$$\text{s. t.} \quad (y_1, y_2) \in \arg\{\min_{y_1, y_2} \quad y_1 + 2y_2 + 1$$

$$\text{s. t.} \quad -25x_1 - 50x_2 - 100x_3 + 20y_1 + 40y_2 \leq 35$$

$$x_1 + 2x_2 + 4x_3 + 2y_1 + 4y_2 \leq 7 \quad\quad (2.22)$$

$$2x_1 + 4x_2 + 8x_3 - y_1 - 2y_2 \leq 14$$

$$-2x_1 - 4x_2 - 8x_3 - 10y_1 + 20y_2 \leq -3\}$$

$$x_1, x_2, x_3 \in \{1, 0\}^3, \quad y_1, y_2 \in \{0, 1\}^2$$

Step 3: The lower level problem is then reformulated as a mp-MILP (2.23), in which the optimization variables of the upper level problem, x_1, x_2 and x_3, are considered as parameters:

$$\min_{y_1, y_2} \quad y_1 + 2y_2 + 1$$

$$\text{s. t.} \quad 25x_1 - 50x_2 - 100x_3 + 20y_1 + 40y_2 \leq 35$$

$$x_1 + 2x_2 + 4x_3 + 2y_1 + 4y_2 \leq 7 \quad\quad (2.23)$$

$$2x_1 + 4x_2 + 8x_3 - y_1 - 2y_2 \leq 14$$

$$-2x_1 - 4x_2 - 8x_3 - 10y_1 + 20y_2 \leq -3$$

$$x_1, x_2, x_3 \in \{1, 0\}^3, y_1, y_2 \in \{0, 1\}^2$$

Step 4: The above problem is then solved using the theory presented in Oberdieck et al. [240] for binary parameters in multiparametric problems, and yields to the optimal parametric solution presented in Table 2.4.

Table 2.4: Lower level problem solution of Example 2.

Critical region	Definition	Objective function	Variables
CR1	$x_2 = 0, x_3 = 0$	2	$y_1 = 1, y_2 = 0$
CR2	$-x_2 - x_3 \leq -1$	1	$y_1 = 0, y_2 = 0$

Steps 5 and 6: The solution obtained is then substituted into the upper level problem to formulate two new single-level ILP problems corresponding to each critical region. Solving these single level problems using CPLEX results to the solution presented in Table 2.5.

After the comparison procedure, the global optimum is found to be -22 with $x = 2$ and $y = 2$.

Table 2.5: Solution of the single level problem formulated in Example 2.

CR	Objective function	Transformed variables	Original variables
CR1	−22	$x_1 = 1, x_2 = 0, x_3 = 0$	$x = 2, y = 2$
CR2	−18	$x_1 = 1, x_2 = 1, x_3 = 1$	$x = 8, y = 1$

2.1.1.3 Example 3: MILP-MILP

Consider the following Type 4^3 class example:

$$\min_{x_{1,2}, y_3} \quad 4x_1 - x_2 + x_3 + 5y_1 - 6y_3$$

$$\text{s. t.} \quad (x_3, y_{1,2}) \in \arg\{\min_{x_3, y_{1,2}} \quad -x_1 + x_2 - 2x_3 - y_1 + 5y_2 + y_3$$

$$\text{s. t.} \quad 6.4x_1 + 7.2x_2 + 2.5x_3 \leq 11.5$$

$$-8x_1 - 4.9x_2 - 3.2x_3 \leq 5 \qquad (2.24)$$

$$3.3x_1 + 4.1x_2 + 0.02x_3 + 4y_1 + 4.5y_2 + 0.5y_3 \leq 1$$

$$-10 \leq x_{1,2} \leq 10\}$$

$$x_1, x_2, x_3 \in \mathbb{R}^3, \quad y_1, y_2, y_3 \in \{0,1\}^3$$

Steps 1 and 2: This example is already bounded and in terms of binary $0 - 1$ variables, therefore, we can directly proceed to Step 3.

Step 3: Considering only the lower level problem, and treating x_1, x_2 and y_3 (upper level variables) as parameters, the lower level problem is transformed to a mp-MILP (2.25).

$$\min_{x_3, y_{1,2}} \quad -x_1 - y_1 + 5y_2 + x_2 - 2x_3 + y_3$$

$$\text{s. t.} \quad 6.4x_1 \leq 11.5 - 7.2x_2 - 2.5x_3$$

$$-8x_1 \leq 5 + 4.9x_2 + 3.2x_3 \qquad (2.25)$$

$$3.3x_1 + 4y_1 + 4.5y_2 \leq 1 - 4.1x_2 - 0.02x_3 - 0.5y_3$$

$$-10 \leq x_{1,2} \leq 10$$

Step 4: Problem (2.25) is then solved using the POP toolbox and the theory presented in Oberdieck et al. [240], and yields the optimal parametric solution shown in Table 2.6.

Step 5: Each solution was then substituted into the upper level problem, resulting into 8 single level linear programming problems corresponding to each critical

---

3 See Table 1.4 for the definition of bilevel mixed-integer problem class types.

Table 2.6: Example 3: Parametric solution of the lower level problem.

CR	Definition	Variables
1	$-0.624x_1 - 0.780x_2 \leq -0.175$ $0.624x_1 + 0.781x_2 \leq 0.198$ $x_1 \leq 10, y_3 = 0$	$x_3 = -165x_1 - 205x_2 + 50$ $y_1 = 0$ $y_2 = 0$
2	$0.624x_1 + 0.781x_2 \leq -0.570$ $-0.624x_1 - 0.780x_2 \leq 0.594$ $x_1 \leq 10, y_3 = 0$	$x_3 = -2.56x_1 - 2.88x_2 + 4.6$ $y_1 = 0$ $y_2 = 0$
3	$-0.626x_1 - 0.780x_2 \leq 0.596$ $0.624x_1 + 0.781x_2 \leq -0.570$ $-0.626x_1 - 0.780x_2 \leq 0.594$ $x_1 \leq 10, y_3 = 0$	$x_3 = -165x_1 - 205x_2 + 50$ $y_1 = 1$ $y_2 = 0$
4	$0.626x_1 + 0.780x_2 \leq -0.596$ $0.044x_1 + 0.999x_2 \leq 4.565$ $-10 \leq x_1 \leq 10$ $-x_2 \leq 10, y_3 = 0$	$x_3 = -2.56x_1 - 2.88x_2 + 4.6$ $y_1 = 1$ $y_2 = 0$
5	$0.044x_1 + 0.999x_2 \leq 4.565$ $0.626x_1 + 0.780x_2 \leq 0.175$ $-0.624x_1 - 0.781x_2 \leq 0.570$ $-10 \leq x_1 \leq 10$ $-x_2 \leq 10, y_3 = 0$	$x_3 = -2.56x_1 - 2.88x_2 + 4.6$ $y_1 = 0$ $y_2 = 0$
6	$0.626x_1 + 0.795x_2 \leq -0.0787$ $-0.624x_1 - 0.781x_2 \leq 0.102$ $x_1 \leq 10, y_3 = 1$	$x_3 = -165x_1 - 205x_2 + 25$ $y_1 = 0$ $y_2 = 0$
7	$0.626x_1 + 0.779x_2 \leq -0.693$ $0.044x_1 + 0.999x_2 \leq 4.565$ $-10 \leq x_1 \leq 10$ $-x_2 \leq 10, y_3 = 1$	$x_3 = -2.56x_1 - 2.88x_2 + 4.6$ $y_1 = 1$ $y_2 = 0$
8	$0.626x_1 + 0.779x_2 \leq 0.0787$ $0.044x_1 + 0.999x_2 \leq 4.565$ $-0.6241x_1 - 0.7814x_2 \leq 0.666$ $-10 \leq x_1 \leq 10$ $-x_2 \leq 10, y_3 = 1$	$x_3 = -2.56x_1 - 2.88x_2 + 4.6$ $y_1 = 0$ $y_2 = 0$

region (2.26):

$$z_1 = \min_{x_{1,2}} \quad -161x_1 - 206x_2 + 50$$
$$\text{s. t.} \quad -0.624x_1 - 0.780x_2 \leq -0.175$$
$$0.624x_1 + 0.781x_2 \leq 0.198$$
$$x_1 \leq 10$$
$$\vdots$$
$$z_8 = \min_{x_{1,2}} \quad 1.44x_1 - 3.88x_2 - 1.4 \tag{2.26}$$
$$\text{s. t.} \quad 0.626x_1 + 0.779x_2 \leq 0.0787$$
$$0.044x_1 + 0.999x_2 \leq 4.565$$
$$-0.6241x_1 - 0.7814x_2 \leq 0.666$$
$$-10 \leq x_1 \leq 10$$
$$-x_2 \leq 10$$

Remark 2.1. Mixed integer linear or quadratic bilevel problems with **all** of the binary variables appearing in the lower level problem will result into pure continuous single-level programming problems at **Step 5** of the algorithm.

Step 6: All 8 linear programming problems (2.26) were solved using the CPLEX solver and their solution is reported in Table 2.7.

Table 2.7: Solution of the single level problems generated in Example 3.

Critical region	Objective	x_1	x_2	x_3	y_1	y_2	y_3
CR1	−38.115	−10	8.243	10.128	0	0	0
CR2	−37.969	−10	7.26	9.291	0	0	0
CR3	173.636	−8.835	6.329	210.306	1	0	0
CR4	−24.457	−7.032	4.879	8.549	1	0	0
CR5	−24.438	−7.020	4.879	8.520	0	0	0
CR6	61.086	−2.736	2.055	80.083	0	0	1
CR7	−25.708	−7.187	4.886	8.928	1	0	1
CR8	−30.704	−7.185	4.886	8.921	0	0	1

After the comparison procedure, the solution with the minimum objective value was chosen as the global solution of the bilevel programming problem (2.25), lying in critical region 1, with $x_1 = -10$, $x_2 = 8.243$, $x_3 = 10.128$ and $y_{1,2,3} = 0$.

2.2 Mixed-integer quadratic bilevel optimization

The algorithm presented before is extended for bilevel mixed-integer quadratic programming problems of the following general form (2.27), belonging to problem

Type 4.[4]

$$
\begin{aligned}
\min_{x_1, y_1} \quad & \left(Q_1^T \omega + c_1\right)^T \omega + c_{c1} \\
\text{s.t.} \quad & A_1 x + B_1 y \le b_1 \\
& (x_2, y_2) \in \arg\Big\{ \min_{x_2, y_2} \quad \left(Q_2^T \omega + c_2\right)^T \omega + c_{c2} \\
& \qquad\qquad\quad \text{s.t.} \quad A_2 x + B_2 y \le b_2 \Big\} \\
& x \in \mathbb{R}^n, \quad y \in \mathbb{Z}^p \\
& x = \begin{bmatrix} x_1^T & x_2^T \end{bmatrix}^T, \quad y = \begin{bmatrix} y_1^T & y_2^T \end{bmatrix}^T, \quad \omega = \begin{bmatrix} x^T & y^T \end{bmatrix}^T
\end{aligned}
\tag{2.27}
$$

where Q_1, c_1, c_{c1}, A_1, B_1, b_1 are constant coefficient matrices in the upper level (leader) problem, and $Q_2 > 0$, c_2, c_{c2}, A_2, B_2, b_2 are constant coefficient matrices in the lower level (follower) problem. It is assumed that upper level optimization variables that appear in the lower level problem, and lower level integer variables are bounded, or their bounds can be derived through the problem constraints.

The main idea and methodology for solving this type of problems follows the methodology proposed in the previous subsection. The algorithm that will be discussed in this section is based on the mp-MIQP algorithm by Oberdieck and Pistikopoulos [199], summarized in Appendix B. The proposed methodology will be first introduced through the general form of the B-MIQP problem (2.27), and then illustrated through two numerical examples.

The first three steps of the B-MIQP algorithm are similar to the first three steps of the B-MILP algorithm. In Step 1, integer and continuous variable bounds are established and in Step 2 integer variables are transformed into binary variables similar to Steps 1 and 2 of the B-MILP algorithm. In Step 3, the lower level problem of the reformulated B-MIQP is transformed into a mp-MIQP problem (2.28), in which the optimization variables of the upper level problem that appear in the lower level problem, x_1 and y_1, are considered as parameters for the lower level problem:

$$
\begin{aligned}
\min_{x_2, y_2} \quad & \left(Q_2^T \omega + c_2\right)^T \omega + c_{c2} \\
\text{s.t.} \quad & A_2 x + B_2 y \le b_2 \\
& x_1^L \le x_1 \le x_1^U
\end{aligned}
\tag{2.28}
$$

The solution of the mp-MIQP problem (2.28), using mp-MIQP algorithms through the POP toolbox, will result to the complete profile of optimal solutions of the lower level problem as explicit functions of the variables of the upper level problem with corresponding critical regions. Because of the mixed-integer terms an exact comparison procedure (minmax, affine, exact) for overlapping critical regions is performed [242]. The quadratic objective function of the lower level problem can therefore make

---

4 See Table 1.4 for the definition of bilevel mixed-integer problem class types.

the final critical regions nonconvex. This is caused during the exact comparison procedure, by the creation of nonlinear inequalities for the definition of the final critical regions (2.29) [199].

$$[x_2, y_2] = \begin{cases} \xi_1 = p_1 + q_1 x_1 + x_1{}^T r_1 x_1, & \psi_1 & \text{if } H_1 x_1 \leq h_1, g_1(x_1) \leq g_1 \\ \xi_2 = p_2 + q_2 x_1 + x_1{}^T r_2 x_1, & \psi_2 & \text{if } H_2 x_1 \leq H_2, g_2(x_1) \leq g_2 \\ \vdots & & \vdots \\ \xi_k = p_k + q_k x_1 + x_1{}^T r_k x_1, & \psi_k & \text{if } H_k x_1 \leq h_k, g_k(x_1) \leq g_k \end{cases} \qquad (2.29)$$

Therefore, in Step 5 we substitute the multiparametric solution into the upper level MIQP problem to formulate single level MIQP or MINLP problems. In Step 6, the single level problems are solved using appropriate mixed-integer linear, quadratic or nonlinear global optimization solvers, and their solutions are compared to select the global optimum solution.

The proposed algorithm in summarized in Table 2.8.

Table 2.8: Multiparametric based algorithm for the solution of B-MIQP problems.

Step 1	Establish integer and continues variable bounds
Step 2	Transform the B-MIQP into a binary B-MIQP problem
Step 3	Recast the lower level as a mp-MIQP, in which the optimization variables of the upper level problem are considered as parameters
Step 4	Solve the resulting mp-MIQP problem to obtain the optimal solution of the lower level as explicit functions of the upper level variables
Step 5	Substitute each multiparametric solution into the upper level problem to formulate k single level MIQP problems
Step 6	Solve all k single level problems using mixed integer quadratic or global algorithms and compare their solutions to select the exact and global optimum

Remark 2.2. This algorithm achieves exact and global optimal solutions when $Q_1 > 0$. For problem cases where this property does not hold, this algorithm is able to achieve approximate global optimum solutions.

2.2.1 Numerical examples

Two B-MIQP numerical examples will be solved to illustrate the use of the proposed algorithm.

2.2.1.1 Example 4: QP-IQP

Consider the following Type 4^5 class example taken from [123]:

$$\min_{x} \quad (x - 2)^2 + (y - 2)^2$$
$$\text{s.t.} \quad \min_{y} \quad y^2$$
$$\text{s.t.} \quad -2x - 2y \le -5$$
$$x - y \le 1 \qquad\qquad\qquad (2.30)$$
$$3x + 2y \le 8$$
$$x \in \mathbb{R}, \quad y \in \{0, 1, 2\}$$

Step 1: Bounds are established for the unbounded continuous variable x (y is already bounded), resulting to $\frac{1}{2} \le x \le \frac{8}{3}$.

Step 2: The problem is reformulated into a 0-1 binary B-MIQP (2.31) by expressing the integer variable y as a linear function of new binary variables y_1 and y_2, $y = y_1 + 2y_2$:

$$\min_{x} \quad (x - 2)^2 + (y_1 + 2y_2 - 2)^2$$
$$\text{s.t.} \quad \min_{y_1, y_2} \quad (y_1 + 2y_2)^2$$
$$\text{s.t.} \quad -2x - 2y_1 - 4y_2 \le -5$$
$$x - y_1 - 2y_2 \le 1 \qquad\qquad\qquad (2.31)$$
$$3x + 2y_1 + 4y_2 \le 8$$
$$x \in \mathbb{R}, \quad y_1, y_2 \in \{0, 1\}^2$$

Step 3: The lower level problem is then reformulated as a mp-MIQP problem (2.32), by considering the upper level optimization variable, x, as a parameter.

$$\min_{y_1, y_2} \quad (y_1 + 2y_2)^2$$
$$\text{s.t.} \quad -2y_1 - 4y_2 \le 2x - 5$$
$$-y_1 - 2y_2 \le -x + 1 \qquad\qquad\qquad (2.32)$$
$$2y_1 + 4y_2 \le -3x + 8$$
$$\frac{1}{2} \le x \le \frac{8}{3}$$

Step 4: The resulting mp-MIQP problem (2.32) is solved using POP toolbox, resulting to the optimal solution presented in Table 2.9.

Step 5: The two solutions were then substituted into the upper level problem, resulting into two single level quadratic programming problems, (2.33) and (2.34), (see

---

5 See Table 1.4 for the definition of bilevel mixed-integer problem class types.

Table 2.9: Lower level problem solution of Example 4.

Critical region	Definition	Objective function	Variables
CR1	$1.5 \leq x \leq 2$	1	$y_1 = 1, y_2 = 0$
CR2	$0.5 \leq x \leq 4/3$	4	$y_1 = 0, y_2 = 1$

Remark 2.1) corresponding to each critical region:

$$z_1 = \min_{x} \quad (x-2)^2 + 1$$
$$\text{s.t.} \quad 1.5 \leq x \leq 2 \tag{2.33}$$

$$z_2 = \min_{x} \quad (x-2)^2$$
$$\text{s.t.} \quad 0.5 \leq x \leq 4/3 \tag{2.34}$$

Step 6: The resulting problems are convex quadratic programming problems therefore CPLEX solver was used for their solution (Table 2.10). After comparison the global solution of the problem was found to be at $x = 4/3$ and $y = 2$ with the objective value of 4/9.

Table 2.10: Solution of the single level problem formulated in Example 4.

Critical region	Objective function	Variables
CR1	5	$x = 2, y = 1$
CR2	4/9	$x = 4/3, y = 2$

2.2.1.2 Example 5: MIQP-MIQP

Consider the following Type 4^6 class example problem:

$$\min_{x_1, x_2, y_3} \quad 4x_1^2 - x_2^2 + 2x_2 + x_3 + 5y_1 + 6y_3$$
$$\text{s.t.} \quad -y_1 - y_2 - y_3 \leq -1$$
$$\min_{x_3, y_1, y_2} \quad 4x_3^2 + y_1^2 + 5y_2 + x_2y_1 - x_2y_2 - 5x_3 - 15y_1 - 16y_2$$
$$\text{s.t.} \quad 6.4x_1 + 7.2x_2 + 2.5x_3 \leq 11.5$$
$$\quad -8x_1 - 4.9x_2 - 3.2x_3 \leq 5$$
$$\quad 3.3x_1 + 4.1x_2 + 0.02x_3 + 4y_1 + 4.5y_2 + 0.5y_3 \leq 1 \tag{2.35}$$
$$\quad -10 \leq x_1 \leq 10$$
$$\quad -10 \leq x_2 \leq 10$$
$$\quad x_1, x_2, x_3 \in \mathbb{R}^3, \quad y_1, y_2, y_3 \in \{0, 1\}^3$$

---

6 See Table 1.4 for the definition of bilevel mixed-integer problem class types.

Steps 1 and 2: The problem is already bounded and in the form of a binary $0-1$ B-MIQP problem, therefore, we can directly proceed to Step 3.

Step 3: The lower level problem is reformulated as a mp-MIQP problem by considering the upper level optimization variables that appear in the lower level (x_1, x_2, y_3) as parameters.

Step 4: The existence of bilinear terms introduces another step for the solution of this problem, as a z-transformation to eliminate those terms is required. This transformation can be done through POP toolbox, and the resulting mp-MIQP problem is solved again using POP toolbox and the theory presented in Oberdieck et al. [240], resulting to the optimal parametric solution presented in Table 2.11 and Figure 2.1.

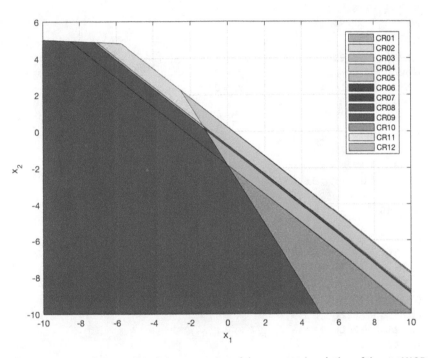

Figure 2.1: Example 5: Graphical representation of the parametric solution of the mp-MIQP problem with 12 critical regions.

Step 5: All 12 critical regions that form the parametric solution were then substituted into the upper level problem, formulating 12 single level MIQP problems corresponding to each critical region.

Step 6: The resulting problems where then solved using CPLEX solver and their solution is presented in Table 2.12. After the comparison procedure, the global optimum was found to be in critical region 11 with an upper level objective function of -1.742.

Table 2.11: Example 5: A part of the multiparametric solution of the lower level problem.

CR	Definition	Variables
1	$0.624x_1 + 0.781x_2 \leq 0.198$ $-0.627x_1 - 0.779x_2 \leq -0.185$ $x_1 \leq 10$	$x_3 = -165x_1 - 205x_2 + 50$ $y_1 = 0$ $y_2 = 0$
2	$0.044x_1 + 0.999x_2 \leq 4.565$ $0.624x_1 + 0.781x_2 \leq 0.198$ $0.853x_1 + 522x_2 \leq -0.959$ $-0.6241x_1 - 0.7814x_2 \leq 0.57$ $-x_1 \leq 10$ $-x_2 \leq 10$	$x_3 = -2.5x_1 - 1.531x_2 - 1.563$ $y_1 = 0$ $y_2 = 0$
3	$0.624x_1 + 0.781x_2 \leq -0.57$ $-0.627x_1 - 0.779x_2 \leq -0.575$ $x_1 \leq 10$	$x_3 = 1.25$ $y_1 = 0$ $y_2 = 0$
4	$-0.853x_1 - 0.5223x_2 \leq 0.959$ $0.627x_1 + 0.779x_2 \leq 0.185$ $-0.624x_1 - 0.781x_2 \leq 0.57$ $x_1 \leq 10$ $-x_2 \leq 10$	$x_3 = 1.25$ $y_1 = 0$ $y_2 = 0$
5	$0.624x_1 + 0.7814x_2 \leq -0.57$ $0.044x_1 + 0.999x_2 \leq 4.564$ $0.853x_1 + 0.522x_2 \leq -0.959$ $-0.624x_1 - 0.781x_2 \leq 0.666$ $-x_1 \leq 10$ $-x_2 \leq 10$	$x_3 = -2.5x_1 - 1.531x_2 - 1.563$ $y_1 = 1$ $y_2 = 0$
n	...	...
11	$0.624x_1 + 0.7814x_2 \leq -1.434$ $-0.627x_1 - 0.779x_2 \leq 1.430$ $x_1 \leq 10$	$x_3 = 1.25$ $y_1 = 0$ $y_2 = 1$
12	$0.627x_1 + 0.779x_2 \leq -0.670$ $-0.853x_1 - 0.522x_2 \leq 0.959$ $-0.624x_1 - 0.781x_2 \leq 1.434$ $x_1 \leq 10$ $-x_2 \leq 10$	$x_3 = 1.25$ $y_1 = 0$ $y_2 = 1$

2.3 Mixed-integer bilevel optimization with right-hand side uncertainty

The presence of uncertainty in bilevel problems has been addressed before for the continuous linear case in Ryu et al. [14] and the continuous quadratic case in Faisca et al. [99], while extensions that cover both mixed integer linear and quadratic cases have been addressed in Avraamidou and Pistikopoulos [91]. In this section of the book, we

Table 2.12: Example 5: Single level solutions.

CR	Variables	Obj. level 1
1	$x_1 = 1.283, x_2 = -0.771, y_3 = 1, x_3 = -3.589, y_1 = 0, y_2 = 0$	3.913
2	$x_1 = -1.328, x_2 = 0.331, y_3 = 1, x_3 = 1.25, y_1 = 0, y_2 = 0$	10.951
3	$x_1 = 0.565, x_2 = -1.193, y_3 = 1, x_3 = 1.25, y_1 = 0, y_2 = 0$	6.790
4	$x_1 = 0.563, x_2 = -1.179, y_3 = 1, x_3 = 1.25, y_1 = 0, y_2 = 0$	6.810
5	$x_1 = -1.180, x_2 = 0.090, y_3 = 0, x_3 = 1.25, y_1 = 1, y_2 = 0$	7.825
6	$x_1 = 0.544, x_2 = -1.298, y_3 = 0, x_3 = 1.25, y_1 = 1, y_2 = 0$	4.372
7	$x_1 = 0.542, x_2 = -1.285, y_3 = 0, x_3 = 1.25, y_1 = 1, y_2 = 0$	4.386
8	$x_1 = 0.530, x_2 = -2.702, y_3 = 0, x_3 = 1.25, y_1 = 1, y_2 = 1$	2.952
9	$x_1 = -0.002, x_2 = -1.834, y_3 = 0, x_3 = 1.25, y_1 = 0, y_2 = 1$	-1.577
10	$x_1 = 0.530, x_2 = -2.702, y_3 = 0, x_3 = 1.25, y_1 = 1, y_2 = 1$	2.952
11	$x_1 = 0.375, x_2 = -2.137, y_3 = 0, x_3 = 1.25, y_1 = 0, y_2 = 1$	-1.742
12	$x_1 = 0.373, x_2 = -2.133, y_3 = 0, x_3 = 1.25, y_1 = 0, y_2 = 1$	-1.740

will discuss the later case, while also considering the uncertainty to be unstructured but bounded, and can appear in one or both optimization levels.

We will address the following bilevel programming problem with right-hand side uncertainty θ:

$$
\begin{aligned}
\min_{x_1, y_1} \quad & \left(Q_1^T \omega + H_{t1}\theta + c_1\right)^T \omega + (Q_{t1}\theta + c_{t1})^T \theta + c_{c1} \\
\text{s.t.} \quad & A_1 x + B_1 y \leq b_1 + F_1 \theta \\
& \min_{x_2, y_2} \quad \left(Q_2^T \omega + H_{t2}\theta + c_2\right)^T \omega + (Q_{t2}\theta + c_{t2})^T \theta + c_{c2} \\
& \text{s.t.} \quad A_2 x + B_2 y \leq b_2 + F_2 \theta \\
& x = \begin{bmatrix} x_1^T & x_2^T \end{bmatrix}^T, \quad x \in \mathbb{R}^n \\
& y = \begin{bmatrix} y_1^T & y_2^T \end{bmatrix}^T, \quad y \in \mathbb{Z}^p \\
& \omega = \begin{bmatrix} x^T & y^T \end{bmatrix}^T \\
& \theta \in \Theta := \left\{\theta \in \mathbb{R}^q \mid M\theta \leq d\right\}
\end{aligned}
$$

(2.36)

For the solution of this problem, we follow the following steps:

Step 1: Similar to the B-MILP and B-MIQP algorithms, integer and continuous variable bounds are established for the variables that appear in the lower level problem.

Step 2: Integer variables are transformed into binary 0-1 variables.

Step 3: The lower level problem is transformed into a mp-MIQP or mp-MILP, considering as parameters both the upper level variables that appear in the lower lever (x_1, y_1), and the uncertainty θ.

Step 4: The resulting multiparametric problems are solved using the POP® toolbox.

Step 5: Each critical region is substituted into the upper level problem to result into k single level multiparametric problems, considering the uncertainty θ as parameters.

Step 6: The resulting k multiparametric problems are solved using the POP® toolbox.

Step 7: All k parametric solutions are combined. For overlapping critical regions, the exact comparison procedure implemented in the POP® toolbox and presented in [199] is used to result to the final exact and global parametric solution of the original bilevel problem.

2.3.1 Numerical example

2.3.1.1 Example 6: mp-MIQP-MILP

Consider the following Type 4^7 class example with right-hand side uncertainty θ:

$$
\begin{aligned}
\min_{x_1, y_3} \quad & 4x_1{}^2 + x_3 y_3 + 5y_1 - 6y_3 - \theta^2 + 2\theta \\
\text{s. t.} \quad & y_1 + y_2 + y_3 \leq 1 \\
& \min_{x_2, y_{1,2}} \quad -x_1 - 2x_2 - y_1 + 5y_2 + \theta \\
& \quad \text{s. t.} \quad 6.4x_1 + 2.5x_2 \leq 11.5 - 7.2\theta \\
& \qquad\qquad\; -8x_1 - 3.2x_2 \leq 5 + 4.9\theta \\
& \qquad\qquad\; 3.3x_1 + 0.02x_2 + 4y_1 + 4.5y_2 \leq 1 - 4.1\theta \\
& \qquad\qquad\; -10 \leq x_1 \leq 10 \\
& \qquad\qquad\; -10 \leq \theta \leq 10 \\
& \quad x_1, x_2 \in \mathbb{R}^2, \quad y_1, y_2, y_3 \in \{0, 1\}^2
\end{aligned}
\tag{2.37}
$$

Steps 1 and 2: The problem is already bounded and in a binary form.

Step 3: The lower level problem is transformed into a mp-MILP problem. Both the upper level variables that appear in the lower level (x_1) and uncertainty (θ) are being treated as parameters for the lower level problem.

Step 4: The problem is then solved using POP toolbox, and yields to the optimal parametric solution presented in Table 2.13.

Step 5: The solutions obtained for every critical region are then substituted into the upper level problem to formulate five new single level mp-MIQP problems. More specifically, the functions of the optimization variables of the lower level, x_2, y_1 and y_2, in terms of the upper level optimization variables, x_1 and θ, are substituted in the upper level problem. The definition of each critical region is added to each new single level problem as a new set of constraints.

---

7 See Table 1.4 for the definition of bilevel mixed-integer problem class types.

Table 2.13: Example 6: Multiparametric solution of the lower level problem.

CR	Definition	Variables
1	$-0.624x_1 - 0.780\theta \leq -0.175$ $0.624x_1 + 0.781\theta \leq 0.198$ $x_1 \leq 10$	$x_2 = -165x_1 - 205\theta + 50$ $y_1 = 0$ $y_2 = 0$
2	$0.624x_1 + 0.781\theta \leq -0.570$ $-0.624x_1 - 0.780\theta \leq 0.594$ $x_1 \leq 10$	$x_2 = -2.56x_1 - 2.88\theta + 4.6$ $y_1 = 0$ $y_2 = 0$
3	$-0.626x_1 - 0.780\theta \leq 0.596$ $0.624x_1 + 0.781\theta \leq -0.570$ $-0.626x_1 - 0.780\theta \leq 0.594$ $x_1 \leq 10$	$x_2 = -165x_1 - 205\theta + 50$ $y_1 = 1$ $y_2 = 0$
4	$0.626x_1 + 0.780\theta \leq -0.596$ $0.044x_1 + 0.999\theta \leq 4.565$ $-10 \leq x_1 \leq 10$ $-\theta \leq 10$	$x_2 = -2.56x_1 - 2.88\theta + 4.6$ $y_1 = 1$ $y_2 = 0$
5	$0.044x_1 + 0.999\theta \leq 4.565$ $0.626x_1 + 0.780\theta \leq 0.175$ $-0.624x_1 - 0.781\theta \leq 0.570$ $-10 \leq x_1 \leq 10$ $-\theta \leq 10$	$x_2 = -2.56x_1 - 2.88\theta + 4.6$ $y_1 = 0$ $y_2 = 0$

Step 6: The five resulting single level problems are in the form of mp-MIQP problems, with the uncertainty θ being a parameter of the single level problems. Therefore, the POP toolbox was used for their solution. Each critical region formed in Step 4 is now divided into smaller regions as another parametric programming problem is solved within the original regions. A summary of the resulting parametric solutions of all five problems is presented in Table 2.14 and Figure 2.2.

Table 2.14: Example 6: A part of the solutions of the single level mp-MIQPs.

CR	Definition	Objective
1.1	$-4.824 \leq \theta \leq 7.733$	$2.136\theta^2 - 408.010\theta - 8.154$
1.2	$7.733 \leq \theta \leq 7.812$	$-\theta^2 - 203\theta - 1406$
...	...	...
5.1	$0.290 \leq \theta \leq 1.241$	$-\theta^2 - 0.880\theta - 2.219$
5.2	$-4.824 \leq \theta \leq 0.290$	$2.096\theta^2 - 2.674\theta - 1.959$
5.3	$-4.882 \leq \theta \leq -4.824$	$0.001\theta^2 + 0.0092\theta + 0.0002$
5.4	$1.2407 \leq \theta \leq 8.7160$	$2.136\theta^2 - 8.661\theta - 2.607$

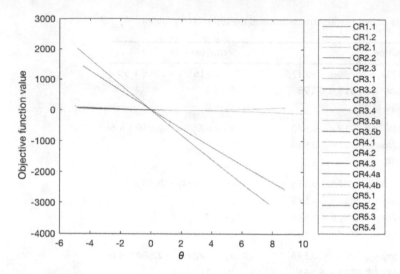

Figure 2.2: Example 6: Graphical representation of the multiparametric solution of the single level mp-MIQP problems.

Step 7: As a last step, the solutions generated from each critical region are compared and the parametric solutions resulting to the minimum objective through the parametric space are chosen as the final solution of the mixed integer bi-level programming problem with uncertainty. Table 2.15 summarizes the final solution of this problem.

Table 2.15: Example 6: Final solution.

CR	Definition	Objective
4.1	$-5.014 \leq \theta \leq -4.882$	$0.011\theta^2 + 0.009\theta + 0.0002$
3.1	$-4.882 \leq \theta \leq -4.840$	$2.136\theta^2 - 2.575\theta + 6.669$
5.3	$-4.840 \leq \theta \leq -4.824$	$0.011\theta^2 + 0.009\theta + 0.0002$
5.2	$-4.824 \leq \theta \leq -0.015$	$2.096\theta^2 - 2.674\theta - 1.959$
1.1	$-0.015 \leq \theta \leq 7.733$	$2.136\theta^2 - 408.010\theta - 8.154$
1.2	$7.733 \leq \theta \leq 7.812$	$-\theta^2 - 203\theta - 1406$
3.4	$7.812 \leq \theta \leq 8.799$	$2.096\theta^2 - 310.374\theta + 4.065$
3.3	$8.799 \leq \theta \leq 8.802$	$-\theta^2 - 203\theta - 701$
4.2	$8.802 \leq \theta \leq 10$	$-\theta^2 - 0.880\theta - 1.095$

2.3.2 Conclusion

This chapter discussed algorithms for the exact global solution of mixed-integer linear and quadratic bilevel problems with integer and continuous variables in both optimization levels. The extension to bilevel problems with right-hand side uncertainty was also discussed. The algorithms presented utilize multiparametric programming to solve the lower level problem as a function of the upper level variables.

The next chapter will discuss the extension of the presented approaches to address more challenging multilevel optimization problems that include more optimization levels, nonlinear terms or multiple optimization problems in a single level.

3 Multilevel optimization

This chapter will discuss the extension of the algorithms presented in Chapter 2, to address more challenging multilevel optimization problems that include (i) more optimization levels, (ii) nonlinear terms, or (iii) multiple optimization problems in a single optimization level.

The first part of this chapter focuses on the solution of mixed-integer linear trilevel optimization problems, while the second part expands the presented approach to address mixed-integer quadratic trilevel optimization problems. The third part extents even more to address certain classes of nonlinear mixed-integer multilevel optimization problems, while the last part considers the case of multiple followers (multi-follower optimization). All algorithms presented in the chapter are explained in detail through step-by-step numerical examples.

3.1 Mixed-integer linear trilevel optimization

When considering a trilevel problem, the feasible set of the third-level problem is parametric in terms of the second- and first-level decision variables, and the second-level problem is parametric in terms of the first-level variables.

Faisca et al. [144] presented an algorithm based on multiparametric programming that can address continuous multilevel programming problems. The approach that we will focus on here is based upon the bilevel mixed-integer linear and quadratic algorithms presented in Chapter 2 [243].

The solution methodology will be first introduced through the general form of the trilevel mixed-integer linear programming (T-MILP) problem (3.1), and then illustrated through 2 numerical examples (Section 3.1.1).

$$
\begin{aligned}
&\min_{x_1, y_1} \quad z(x, y) = c_1{}^T x + d_1{}^T y \\
&\text{s.t.} \quad A_1 x + B_1 y \leq b_1 \\
&\qquad \min_{x_2, y_2} \quad u(x, y) = c_2{}^T x + d_2{}^T y \\
&\qquad \text{s.t.} \quad A_2 x + B_2 y \leq b_2 \\
&\qquad\qquad \min_{x_3, y_3} \quad v(x, y) = c_3{}^T x + d_3{}^T y \\
&\qquad\qquad \text{s.t.} \quad A_3 x + B_3 y \leq b_3 \\
&\qquad x = \begin{bmatrix} x_1^T & x_2^T & x_3^T \end{bmatrix}^T, \quad y = \begin{bmatrix} y_1^T & y_2^T & y_3^T \end{bmatrix}^T \\
&\qquad x \in \mathbb{R}^n, \quad y \in \mathbb{Z}^p
\end{aligned}
\tag{3.1}
$$

https://doi.org/10.1515/9783110760316-003

where $c_i \in \mathbb{R}^n$, $d_i \in \mathbb{R}^p$, $A_i \in \mathbb{R}^{m_i \times n}$, $B_i \in \mathbb{R}^{m_i \times p}$, $b_i \in \mathbb{R}^{m_i}$, x_1 and x_2 are compact (closed and bounded), x is a vector of the continuous problem variables and y is a vector of the discrete problem variables. The subscript numbers (1, 2 and 3) indicate the optimization level the constant coefficient matrices and the decision variables belong to.

The first step of the methodology is to establish bounds for all the integer and continuous variables, and transform the T-MILP problem into a binary T-MILP problem by expressing integer variables in terms of binary 0-1 variables, following the method presented in Floudas [241] Section 6.2.1, Remark 1.

As a second step, the third-level problem of the binary T-MILP, is transformed into a multiparametric mixed-integer linear programming (mp-MILP) problem (3.2), by considering the optimization variables of the first- and second-level problems, x_1, y_1, x_2, and y_2, as parameters.

$$\min_{x_3, y_3} \quad v(x, y) = c_3^T x + d_3^T y$$
$$\text{s.t.} \quad A_3 x + B_3 y \le b_3 \tag{3.2}$$
$$x^L \le x \le x^U$$

The solution of (3.2) using multiparametric programming theory for mp-MILP problems [198] and the theory for binary parameters in multiparametric programming problems [240] embedded in POP® toolbox [242], provides the complete profile of optimal solutions of the third-level problem as explicit functions of the decision variables of the first- and second-level problems (3.3):

$$x_3 = \begin{cases} \xi_1 = p_1 + q_1 [x_1^T \ y_1^T \ x_2^T \ y_2^T]^T & \text{if } H_1 [x_1^T \ y_1^T \ x_2^T \ y_2^T]^T \le h_1, y_3 = r_1 \\ \xi_2 = p_2 + q_2 [x_1^T \ y_1^T \ x_2^T \ y_2^T]^T & \text{if } H_2 [x_1^T \ y_1^T \ x_2^T \ y_2^T]^T \le h_2, y_3 = r_2 \\ \quad \vdots & \qquad \vdots \\ \xi_k = p_k + q_k [x_1^T \ y_1^T \ x_2^T \ y_2^T]^T & \text{if } H_k [x_1^T \ y_1^T \ x_2^T \ y_2^T]^T \le h_k, y_3 = r_k \end{cases} \tag{3.3}$$

where ξ_i is the affine function of the third level continuous variables in terms of the first- and second-level decision variables, $H_i [x_1^T \ y_1^T \ x_2^T \ y_2^T]^T \le h_i$, $y_3 = r_i$ is referred to as critical region i, CR_i and k denotes the number of computed critical regions.

The next step is to recast the second-level problem into k mp-MILP problems (3.4), by considering the optimization variables of the first-level problem, x_1, y_1, as parameters, substituting in the corresponding functions ξ_i of x_3 and y_3 and adding the corresponding critical region definitions as a new set of constraints for each problem:

$$u_1 = \min_{x_2, y_2} \quad c_2^T x + d_2^T y$$
$$\text{s. t.} \quad A_2 x + B_2 y \le b_2$$
$$H_1 [x_1^T \ y_1^T \ x_2^T \ y_2^T]^T \le h_1$$
$$x_3 = \xi_1, \quad y_3 = r_1$$

$$u_2 = \min_{x_2, y_2} \quad c_2^T x + d_2^T y$$
$$\text{s. t.} \quad A_2 x + B_2 y \le b_2$$
$$H_2 [x_1^T \ y_1^T \ x_2^T \ y_2^T]^T \le h_2 \tag{3.4}$$
$$x_3 = \xi_2, \quad y_3 = r_2$$
$$\vdots$$

$$u_k = \min_{x_2, y_2} \quad c_2^T x + d_2^T y$$
$$\text{s. t.} \quad A_2 x + B_2 y \le b_2$$
$$H_k [x_1^T \ y_1^T \ x_2^T \ y_2^T]^T \le h_k$$
$$x_3 = \xi_k, \quad y_3 = r_k$$

The above multiparametric problems (3.4) are independent of each other, therefore, it is possible to use parallel programming to solve them simultaneously. This problems are again solved using POP® toolbox, providing the complete profile of optimal solutions of the second-level problem constraint by the optimality of the third-level problem, as explicit functions of the decision variables of the first-level problem (3.5).

$$x_2 = \begin{cases} \psi_{1,1} = m_{1,1} + n_{1,1} [x_1^T \ y_1^T]^T & \text{if } G_{1,1} [x_1^T \ y_1^T]^T \le g_{1,1}, y_2 = s_{1,1} \\ \psi_{1,2} = m_{1,2} + n_{1,2} [x_1^T \ y_1^T]^T & \text{if } G_{1,2} [x_1^T \ y_1^T]^T \le g_{1,2}, y_2 = s_{1,2} \\ \vdots & \vdots \\ \psi_{1,l_1} = m_{1,l_1} + n_{1,l_1} [x_1^T \ y_1^T]^T & \text{if } G_{1,l_1} [x_1^T \ y_1^T]^T \le g_{1,l_1}, y_2 = s_{1,l_1} \\ \psi_{2,1} = m_{2,1} + n_{2,1} [x_1^T \ y_1^T]^T & \text{if } G_{2,1} [x_1^T \ y_1^T]^T \le g_{2,1}, y_2 = s_{2,1} \\ \psi_{2,2} = m_{2,2} + n_{2,2} [x_1^T \ y_1^T]^T & \text{if } G_{2,2} [x_1^T \ y_1^T]^T \le g_{2,2}, y_2 = s_{2,2} \\ \vdots & \vdots \\ \psi_{2,l_2} = m_{2,l_2} + n_{2,l_2} [x_1^T \ y_1^T]^T & \text{if } G_{2,l_2} [x_1^T \ y_1^T]^T \le g_{2,l_2}, y_2 = s_{2,l_2} \\ \vdots & \vdots \\ \psi_{k,1} = m_{k,1} + n_{k,1} [x_1^T \ y_1^T]^T & \text{if } G_{k,1} [x_1^T \ y_1^T]^T \le g_{k,1}, y_2 = s_{k,1} \\ \psi_{k,2} = m_{k,2} + n_{k,2} [x_1^T \ y_1^T]^T & \text{if } G_{k,2} [x_1^T \ y_1^T]^T \le g_{k,2}, y_2 = s_{k,2} \\ \vdots & \vdots \\ \psi_{k,l_k} = m_{k,l_k} + n_{k,l_k} [x_1^T \ y_1^T]^T & \text{if } G_{k,l_k} [x_1^T \ y_1^T]^T \le g_{k,l_k}, y_2 = s_{k,l_k} \end{cases} \tag{3.5}$$

where $G_{i,j}[x_1^T y_1^T]^T \leq g_{i,j}, y_2 = s_{i,j}$ is referred to as the critical region j that resulted from the critical region i of the multiparametric solution of the third-level problem ($CR_{i,j}$), $\psi_{i,j}$ is the explicit affine function of the second-level continuous variables in terms of the first-level decision variables, in $CR_{i,j}$, and l_i denotes the number of computed second-level critical regions from third-level critical region i.

The next step of the algorithm is to transform the first-level problem into single-level deterministic mixed-integer programming problems (3.6) corresponding to each $CR_{i,j}$. This can be done by substituting into the first-level problem the affine functions expressing third-level decision variables in terms of second- and first-level decision variables, ξ_i, and the affine functions expressing second-level decision variables in terms of first-level decision variables, $\psi_{i,j}$, eliminating in this way all second- and third-level variables form the first-level problem. Moreover, the critical region definitions are added to the corresponding single-level problems as an additional set of constraints:

$$
\begin{aligned}
z_{1,1} = \min_{x_1, y_1} \quad & c_1^T x + d_1^T y \\
\text{s.t.} \quad & A_1 x + B_1 y \leq b_1 \\
& G_{1,1}[x_1^T \quad y_1^T]^T \leq g_{1,1} \\
& x_2 = \psi_{1,1}, \quad y_2 = s_{1,1} \\
& x_3 = \xi_1, \quad y_3 = r_1 \\
z_{1,2} = \min_{x_1, y_1} \quad & c_1^T x + d_1^T y \\
\text{s.t.} \quad & A_1 x + B_1 y \leq b_1 \\
& G_{1,2}[x_1^T \quad y_1^T]^T \leq g_{1,2} \\
& x_2 = \psi_{1,2}, \quad y_2 = s_{1,2} \\
& x_3 = \xi_1, \quad y_3 = r_1 \\
& \vdots \\
z_{k,l_k} = \min_{x_1, y_1} \quad & c_1^T x + d_1^T y \\
\text{s.t.} \quad & A_1 x + B_1 y \leq b_1 \\
& G_{k,l_k}[x_1^T \quad y_1^T]^T \leq g_{k,l_k} \\
& x_2 = \psi_{k,l_k}, \quad y_2 = s_{k,l_k} \\
& x_3 = \xi_k, \quad y_3 = r_k
\end{aligned}
\tag{3.6}
$$

The single-level deterministic programming problems (3.6) are, similar to problems (3.4), independent of each other, making it possible to use parallel programming to solve them simultaneously.

The single-level MILP problems (3.6) are solved using CPLEX® and their solutions correspond to different local optimal solutions of the original trilevel problem.

The final step of the algorithm is a comparison procedure to select the global optimum solution of the trilevel problem. This procedure is performed by solving the mixed-integer linear programming problem (3.7).

$$z^* = \min_{\alpha, y} \quad \alpha$$

$$\text{s. t.} \quad \alpha = \sum_{i,j} y_{i,j} z_{i,j} \qquad\qquad [C1]$$

$$\sum_{i,j} y_{i,j} = 1 \qquad\qquad [C2] \qquad\qquad (3.7)$$

$$y_{i,j} u_{i,j} \le y_{i,j} u_{p,q} \quad \forall i, j, p \neq i, q \quad [C3]$$

$$y_{i,j} \in \{0, 1\}$$

where z^* is the exact global optimum of the original trilevel MILP programming problem, $y_{i,j}$ are binary variables corresponding to each $CR_{i,j}$, $z_{i,j}$ are the objective function values of the first-level problem obtained when solving problems (3.6), and u_i are the objective function values of the second-level problem obtained when solving problems (3.4). Constraint [C2] makes sure that only one $y_{i,j}$ is equal to 1 and the rest are equal to zero (i. e., only one solution is selected), while constraint [C3] makes sure that the selected solution results to an optimal solution in the second level. When $y_{i,j} = 1$, then the optimal solution lies in $CR_{i,j}$ and CR_i.

The discussed algorithm is summarized in 8 simple steps in Table 3.1.

Table 3.1: Multiparametric based algorithm for the solution of T-MILP problems.

Step 1	Establish integer and continuous variable bounds, and transform the T-MILP into a binary T-MILP
Step 2	Recast the third level problem as a mp-MILP, in which the optimization variables of the second- and first-level problems are considered as parameters
Step 3	Solve the resulting mp-MILP problem to obtain the optimal solution of the lower level problem as explicit functions of the second- and first- level decision variables
Step 4	Substitute each multiparametric solution into the second-level problem to formulate k mp-MILP problems, considering the first-level decision variables as parameters
Step 5	Solve the resulting k mp-MILP problems to obtain the optimal solution of the second-level problem as explicit functions of the first-level decision variables
Step 6	Substitute each multiparametric solution into the first-level problem to formulate single-level MILP problems
Step 7	Solve all single-level problems using CPLEX® MILP solver
Step 8	Solve the comparison optimization problem (2.6) to select the exact and global optimum solution

3.1.1 Numerical examples

Two numerical examples are presented to illustrate the proposed algorithm. All computations were carried out on a 2-core machine with an Intel Core i7 at 3.1 GHz and 16 GB of RAM, MATLAB R2016a, and IBM ILOG CPLEX Optimization Studio 12.6.3.

3.1.1.1 Example 7: Trilevel continuous linear programming problem

Consider the following trilevel continuous linear programming problem (3.8) from Anandalingam [244] (also illustrated and used in other publications including [137, 145, 245]):

$$
\min_{x_1} \quad z = -7x_1 - 3x_2 + 4x_3
$$
$$
\text{s. t.} \quad \min_{x_2} \quad u = -x_2
$$
$$
\text{s. t.} \quad \min_{x_3} \quad v = -x_3
$$
$$
\text{s. t.} \quad
\begin{aligned}
& x_1 + x_2 + x_3 \le 3 \\
& x_1 + x_2 - x_3 \le 1 \\
& x_1 + x_2 + x_3 \ge 1 \\
& -x_1 + x_2 + x_3 \le 1 \\
& x_3 \le 0.5 \\
& x_1, x_2, x_3 \ge 0
\end{aligned}
\tag{3.8}
$$
$$
x_1, x_2, x_3 \in \mathbb{R}^3
$$

Step 1: Continuous variable bounds are established, resulting to the inequalities in (3.9):

$$
\begin{aligned}
0 &\le x_1 \le 1.5 \\
0 &\le x_2 \le 1 \\
0 &\le x_3 \le 0.5
\end{aligned}
\tag{3.9}
$$

Step 2: The third-level problem is reformulated as a mp-LP (3.10), in which the decision variable of the first-level problem, x_1, and the decision variable of the second-level problem, x_2, are considered as parameters:

$$
\min_{x_3} \quad v = -x_3
$$
$$
\text{s. t.} \quad
\begin{aligned}
& x_3 \le 3 - x_1 - x_2 \\
& -x_3 \le 1 - x_1 - x_2 \\
& -x_3 \le -1 + x_1 + x_2 \\
& x_3 \le 1 + x_1 - x_2 \\
& 0 \le x_1 \le 1.5 \\
& 0 \le x_2 \le 1 \\
& 0 \le x_3 \le 0.5
\end{aligned}
\tag{3.10}
$$

Step 3: Problem (3.10) is solved using POP® toolbox [242] and yields the optimal parametric solution given in Table 3.2 and illustrated in Figures 3.1 and 3.2.

Table 3.2: Example 1: Multiparametric solution of the third level mp-LP problem.

Critical region	Definition	3rd level objective	3rd level variables
CR_1	$x_1 \geq 0$ $x_2 \leq 1$ $x_1 - x_2 \leq -0.5$	$v = -x_1 + x_2 - 1$	$x_3 = x_1 - x_2 + 1$
CR_2	$x_2 \geq 0$ $x_1 + x_2 \leq 1.5$ $-x_1 - x_2 \leq -0.5$ $-x_1 + x_2 \leq 0.5$	$v = -0.5$	$x_3 = 0.5$

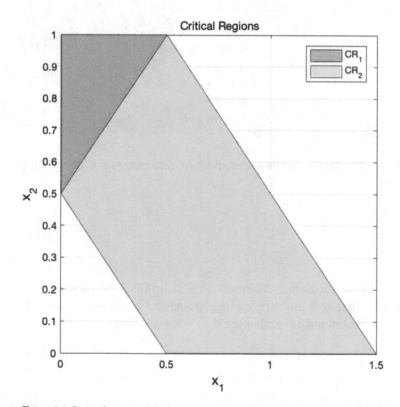

Figure 3.1: Example 7: Graphical representation of the critical regions forming the multiparametric solution of the third level mp-LP problem.

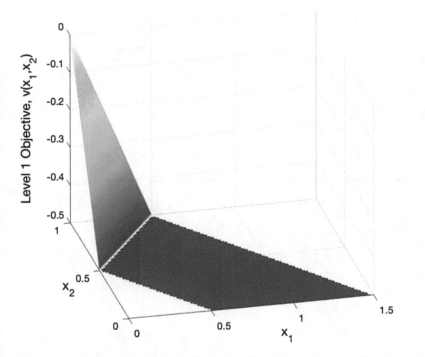

Figure 3.2: Example 7: Graphical representation of the objective function of level 1 as a function of the parameters x_1 and x_2.

Step 4: The solution in Table 3.2 is then substituted into the second-level problem to formulate two new mp-LP problems ((3.11) and (3.12)) corresponding to each critical region, by considering the decision variable of the first level, x_1, as a parameter, adding the critical region definition as a new set of constraints and substituting x_3 with its corresponding affine function in terms of x_1 and x_2:

$$\min_{x_2} \quad u_1 = -x_2$$
$$\text{s.t.} \quad 0 \leq x_1$$
$$x_2 \leq 1$$
$$-x_2 \leq -0.5 - x_1$$
$$\tag{3.11}$$

$$\min_{x_2} \quad u_2 = -x_2$$
$$\text{s.t.} \quad 0 \leq x_2$$
$$x_1 + x_2 \leq 1.5$$
$$-x_1 - x_2 \leq -0.5$$
$$-x_1 + x_2 \leq 0.5$$
$$\tag{3.12}$$

Step 5: The two mp-LP problems (3.11) and (3.12) are then solved using POP® toolbox [242] and resulted to the parametric solution given in Table 3.3 and illustrated in Figure 3.3.

Table 3.3: Example 7: Multiparametric solution of the second-level problem.

Critical region	Definition	2nd level objective	2nd level variables
$CR_{1,1}$	$x_1 \geq 0$ $x_1 \leq 0.5$	$u_{1,1} = -1$	$x_2 = 1$
$CR_{2,1}$	$x_1 \geq 0$ $x_2 \leq 0.5$	$u_{2,1} = -x_1 - 0.5$	$x_2 = x_1 + 0.5$
$CR_{2,2}$	$x_1 \geq 0.5$ $x_1 \leq 1.5$	$u_{2,2} = x_1 - 1.5$	$x_2 = -x_1 + 1.5$

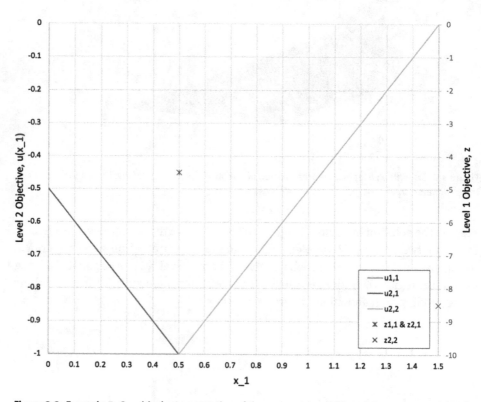

Figure 3.3: Example 7: Graphical representation of the parametric solution of the two second-level mp-LP problems, and the deterministic solution of the three first-level LP problems.

Step 6: The parametric solutions obtained (Table 3.3) are then used to formulate three single-level deterministic linear programming problems ((3.13), (3.14) and (3.15)), each corresponding to one critical region. The critical region definition is added to the first-level problem as a new set of constraints and the variables x_2 and x_3 are substituted by their derived affine expressions, resulting into three linear programming problems that involve only x_1:

$$\min_{x_1} \quad z_{1,1} = -7x_1 - 3(1) + 4(x_1)2$$
$$\text{s.t.} \quad 0 \le x_1 \le 0.5 \tag{3.13}$$

$$\min_{x_1} \quad z_{2,1} = -7x_1 - 3(x_1 + 0.5) + 4(0.5)$$
$$\text{s.t.} \quad 0 \le x_1 \le 0.5 \tag{3.14}$$

$$\min_{x_1} \quad z_{2,2} = -7x_1 - 3(-x_1 + 1.5) + 4(0.5)$$
$$\text{s.t.} \quad 0.5 \le x_1 \le 1.5 \tag{3.15}$$

Step 7: Problems (3.13) to (3.15) are solved using CPLEX® linear programming solver, and the resulting solutions are presented in Table 3.4 and illustrated in Figure 3.3.

Table 3.4: Example 7: First-level problem solutions.

Critical region	Objectives	Decision variables
$CR_{1,1}$	$z_{1,1} = -4.5, u_{1,1} = -1, v = -0.5$	$x_1 = 0.5, x_2 = 1, x_3 = 0.5$
$CR_{2,1}$	$z_{2,1} = -4.5, u_{2,1} = -1, v = -0.5$	$x_1 = 0.5, x_2 = 1, x_3 = 0.5$
$CR_{2,2}$	$z_{2,2} = -8.5, u_{2,2} = 0, v = -0.5$	$x_1 = 1.5, x_2 = 0, x_3 = 0.5$

Step 8: Using the information from Tables 3.2, 3.3 and 3.4, the comparison optimization problem (3.7) is solved to select the exact and global optimum of the problem, that was found to be lying in $CR_{2,2}$. The optimal decision variables are $x_1 = 1.5$, $x_2 = 0$ and $x_3 = 0.5$, with the first level objective $z_{2,2} = -8.5$, the second-level objective $u_{2,2} = 0$, and the third-level objective $v = -0.5$.

The computational performance of the algorithm for this example is presented in Table 3.5.

Table 3.5: Example 7 computational performance.

Problem	Problem type	Number of problems	CPU time (s)
Third level	mp-LP	1	0.061992
Second level	mp-LP	2	0.094649
First level	LP	3	0.029097
Comparison	MILP	1	0.017443

3.1.1.2 Example 8: Trilevel mixed-integer linear programming problem

Consider the following T-MILP problem (3.16) of class Type 3, (i. e., both integer and continuous variables appear in all optimization levels):

$$\min_{x_1,y_1} \quad z = -7x_1 + y_1 - 2x_2$$

$$\text{s. t.} \quad \min_{x_2,y_2} \quad u = -x_1 - 3y_1 - x_2 - 3y_2 + x_3 + 5y_3 + 2y_4$$

$$\text{s. t.} \quad \min_{x_3,y_3,y_4} \quad v = 2x_1 + y_1 + 2x_2 + y_2 - 4x_3 + 2y_3 + 10y_4$$

$$\text{s. t.} \quad 6.4x_1 + 7.2x_2 + 2.5x_3 \le 11.5$$

$$- 8x_1 - 4.9x_2 - 3.2x_3 \le 5 \tag{3.16}$$

$$3.3x_1 + 4.1x_2 + 0.02x_3 + 0.2y_1 + 0.8y_2 + 4y_3 + 4.5y_4 \le 1$$

$$y_1 + y_2 + y_3 + y_4 \ge 1$$

$$- 10 \le x_1, x_2 \le 10$$

$$x_1, x_2, x_3 \in \mathbb{R}^3, \quad y_1, y_2, y_3, y_4 \in \{0, 1\}^4$$

Step 1: Both binary and continuous variables are bounded, therefore, we can continue to Step 2.

Step 2: The third-level problem is reformulated as a mp-MILP problem (3.17), in which both continuous and binary decision variables of the first-level problem, x_1, y_1, and the second-level problem, x_2, y_2, are considered as parameters:

$$\min_{x_3,y_3,y_4} \quad v = 2x_1 + y_1 + 2x_2 + y_2 - 4x_3 + 2y_3 + 10y_4$$

$$\text{s. t.} \quad 6.4x_1 + 7.2x_2 + 2.5x_3 \le 11.5$$

$$-8x_1 - 4.9x_2 - 3.2x_3 \le 5$$

$$3.3x_1 + 4.1x_2 + 0.02x_3 + 0.2y_1 + 0.8y_2 + 4y_3 + 4.5y_4 \le 1 \tag{3.17}$$

$$y_1 + y_2 + y_3 + y_4 \ge 1$$

$$-10 \le x_1, x_2 \le 10$$

$$x_1, x_2, x_3 \in \mathbb{R}^3, \quad y_1, y_2, y_3, y_4 \in \{0, 1\}^4$$

Step 3: Problem (3.17) is solved using POP® toolbox [242] and yield the optimal multiparametric solution presented in Table 3.6.

Step 4: The multiparametric solution in Table 3.6 is substituted into the second-level problem, to formulate 6 new mp-MILP problems, each corresponding to a critical region of the third-level solution. The decision variables of the first- level problem, both continuous and binary (x_1, y_1), are considered as parameters, the critical region definitions are added as a new set of constraints and the explicit function of the third-level variables (x_3, y_3, y_4) in terms of the second- and first- level variables are substituted in the resulting mp-MILP problem. The reformulated second-level problem corresponding to critical region CR_1 is presented as (3.18). Similar mp-MILPs are

Table 3.6: Example 8: multiparametric solution of the third-level problem.

CR	Definition	3rd level obj.	3rd level var.
CR_1	$-y_1 - y_2 \leq -1$ $0.0444x_1 + 0.999x_2 \leq 4.5645$ $0.6187x_1 + 0.0381y_1 + 0.7698x_2$ $\quad + 0.1523y_2 \leq -0.5888$ $-10 \leq x_1 \leq 10$ $-x_2 \leq 10$ $y_1, y_2 \in \{0, 1\}$	$v = 12.24x_1 + y_1$ $\quad + 13.52x_2 + y_2 - 18.4$	$x_3 = -2.56x_1 - 2.88x_2$ $\quad + 4.6$ $y_3 = 0$ $y_4 = 0$
CR_2	$y_1 + y_2 \leq 1$ $0.0444x_1 + 0.999x_2 \leq 4.5645$ $0.6187x_1 + 0.0381y_1 + 0.7698x_2$ $\quad + 0.1523y_2 \leq -0.5888$ $-10 \leq x_1 \leq 10$ $-x_2 \leq 10$ $y_1, y_2 \in \{0, 1\}$	$v = 12.24x_1 + y_1$ $\quad + 13.52x_2 + y_2 - 16.4$	$x_3 = -2.56x_1 - 2.88x_2$ $\quad + 4.6$ $y_3 = 1$ $y_4 = 0$
CR_3	$-y_1 - y_2 \leq -1$ $0.6164x_1 + 0.0377y_1 + 0.7718x_2$ $\quad + 0.1517y_2 \leq -0.5630$ $-0.6187x_1 - 0.0381y_1 - 0.7698x_2$ $\quad - 0.1523y_2 \leq 0.5888$ $x_1 \leq 10$ $y_1, y_2 \in \{0, 1\}$	$v = 12.24x_1 + y_1$ $\quad + 13.52x_2 + y_2 - 18.4$	$x_3 = -2.56x_1 - 2.88x_2$ $\quad + 4.6$ $y_3 = 0$ $y_4 = 0$
CR_4	$y_1 + y_2 \leq 1$ $0.6164x_1 + 0.0377y_1 + 0.7718x_2$ $\quad + 0.1517y_2 \leq -0.5630$ $-0.6187x_1 - 0.0381y_1 - 0.7698x_2$ $\quad - 0.1523y_2 \leq 0.5888$ $x_1 \leq 10$ $y_1, y_2 \in \{0, 1\}$	$v = 662x_1 + 41y_1 + 822x_2$ $\quad + 161y_2 + 602$	$x_3 = -165x_1 - 10y_1$ $\quad - 205x_2 - 40y_2$ $\quad - 150$ $y_3 = 1$ $y_4 = 0$
CR_5	$-y_1 - y_2 \leq -1$ $-0.6164x_1 - 0.0377y_1 - 0.7718x_2$ $\quad - 0.1517y_2 \leq 0.5630$ $0.6187x_1 + 0.0381y_1 + 0.7698x_2$ $\quad + 0.1523y_2 \leq 0.1729$ $0.0444x_1 + 0.999x_2 \leq 4.5645$ $x_1 \leq 10$ $y_1, y_2 \in \{0, 1\}$	$v = 12.24x_1 + y_1 + 13.52x_2$ $\quad + y_2 - 18.4$	$x_3 = -2.56x_1 - 2.88x_2$ $\quad + 4.6$ $y_3 = 0$ $y_4 = 0$
CR_6	$-y_1 - y_2 \leq -1$ $0.6164x_1 + 0.0377y_1 + 0.7718x_2$ $\quad + 0.1517y_2 \leq 0.1956$ $-0.6187x_1 - 0.0381y_1 - 0.7698x_2$ $\quad - 0.1523y_2 \leq -0.1729$ $x_1 \leq 10$ $y_1, y_2 \in \{0, 1\}$	$v = 662x_1 + 41y_1 + 822x_2$ $\quad + 161y_2 - 200$	$x_3 = -165x_1 - 10y_1$ $\quad - 205x_2 - 40y_2$ $\quad + 50$ $y_3 = 0$ $y_4 = 0$

formulated for the rest of the critical regions of the third-level problem:

$$\min_{x_2, y_2} \quad u_1 = -x_1 - 3y_1 - x_2 - 3y_2 + (-2.56x_1 - 2.88x_2 + 4.6) + 5(0) + 2(0)$$

$$\text{s. t.} \quad -y_1 - y_2 \leq -1$$
$$0.0444x_1 + 0.999x_2 \leq 4.5645$$
$$0.6187x_1 + 0.0381y_1 + 0.7698x_2 + 0.1523y_2 \leq -0.5888 \tag{3.18}$$
$$-10 \leq x_1 \leq 10$$
$$-x_2 \leq 10$$
$$y_1, y_2 \in \{0, 1\}^2$$

Step 5: The six mp-MILP problems created in Step 4 are independent of each other and can be solved simultaneously. They are again solved using the POP® toolbox [242], resulting into nine critical regions. Their multiparametric solution is presented in Table 3.7.

Step 6: The parametric solutions of the second-level problem obtained (Table 3.7) are then used to formulate nine single-level deterministic MILP problems, as reformulations of the original trilevel problem. The critical region definitions are added as a new set of constraints to the first-level problem. Also, the second- and first-level variables, x_2, y_2, x_3, y_3, y_4, are substituted by their derived affine functions in terms of the first-level variables, resulting into deterministic MILP problems containing only first-level variables, x_1, y_1. The resulting deterministic problems are independent of each other and can be solved simultaneously. The single-level problems resulting from $CR_{1,1}$ (3.19) and $CR_{6,1}$ (3.20) are presented below:

$$\min_{x_1, y_1} \quad z_{1,1} = -7x_1 + y_1 - 2(-0.0445x_1 - 4.5690)$$

$$\text{s. t.} \quad 0.9979x_1 + 0.065y_1 \leq -7.2710 \tag{3.19}$$
$$-x_1 \leq 10$$
$$y_1 \in \{0, 1\}$$

$$\min_{x_1, y_1} \quad z_{6,1} = -7x_1 + y_1 - 2(-0.7986x_1 - 0.0491y_1 + 0.0568)$$

$$\text{s. t.} \quad -0.9979x_1 - 0.065y_1 \leq 5.9704 \tag{3.20}$$
$$x_1 \leq 10$$
$$y_1 \in \{0, 1\}$$

Step 7: The nine deterministic problems are solved using CPLEX® mixed-integer linear programming solver. The resulting solutions are presented in Table 3.8.

Step 8: Using the information from Tables 3.6, 3.7 and 3.8, the comparison optimization problem (3.7) is solved to select the exact and global optimum of the trilevel

Table 3.7: Example 8: Multiparametric solution of the second-level problem.

CR	Definition	2nd level objective	2nd level var.
$CR_{1,1}$	$0.9979x_1$ $+ 0.065y_1 \leq -7.2710$ $-x_1 \leq 10$ $y_1 \in \{0,1\}$	$u_{1,1} = -3.3874x_1$ $- 3y_1 - 16.1279$	$x_2 = -0.0445x_1$ $- 4.5690$ $y_2 = 1$
$CR_{1,2}$	$-0.9979x_1$ $- 0.065y_1 \leq 7.2710$ $x_1 \leq 10$ $y_1 \in \{0,1\}$	$u_{1,2} = -0.4417x_1$ $- 2.8080y_1 - 5.3356$	$x_2 = -0.8037x_1$ $- 0.0495y_1 - 0.9628$ $y_2 = 1$
$CR_{2,1}$	$0.9979x_1$ $+ 0.065y_1 \leq -7.0109$ $-x_1 \leq 10$ $y_1 \in \{0,1\}$	$u_{2,1} = -3.3874x_1$ $- 3y_1 - 8.1279$	$x_2 = -0.0445x_1$ $- 4.5690$ $y_2 = 0$
$CR_{2,2}$	$-0.9979x_1$ $- 0.065y_1 \leq 7.0109$ $x_1 \leq 10$ $y_1 \in \{0,1\}$	$u_{2,2} = 0.5399x_1$ $- 3.3780y_1 - 12.5959$	$x_2 = -0.8037x_1$ $- 0.0495y_1 - 0.7649$ $y_2 = 0$
$CR_{3,1}$	$-0.9979x_1$ $- 0.065y_1 \leq 7.2710$ $x_1 \leq 10$ $y_1 \in \{0,1\}$	$u_{3,1} = 0.4612x_1$ $- 2.8093y_1 + 5.1934$	$x_2 = -0.7986x_1$ $- 0.0491y_1 - 0.9261$ $y_2 = 1$
$CR_{4,1}$	$-0.9979x_1$ $- 0.065y_1 \leq 7.0109$ $x_1 \leq 10$ $y_1 \in \{0,1\}$	$u_{4,1} = -1.4784x_1$ $- 2.8756y_1 + 5.2841$	$x_2 = -0.7986x_1$ $- 0.0491y_1 - 0.7295$ $y_2 = 0$
$CR_{5,1}$	$0.9979x_1$ $- 0.065y_1 \leq -5.9704$ $-0.9979x_1$ $- 0.065y_1 \leq 7.2710$ $y_1 \in \{0,1\}$	$u_{5,1} = -3.3874x_1$ $- 3y_1 - 16.1279$	$x_2 = -0.0445x_1$ $4.5690y_2 = 1$
$CR_{5,2}$	$-0.9979x_1$ $- 0.065y_1 \leq 5.9704$ $x_1 \leq 10$ $y_1 \in \{0,1\}$	$u_{5,2} = -0.4417x_1$ $- 2.8080y_1 + 1.4963$	$x_2 = -0.8037x_1$ $- 0.0495y_1 + 0.0267$ $y_2 = 1$
$CR_{6,1}$	$-0.9979x_1$ $- 0.065y_1 \leq 5.9704$ $x_1 \leq 10$ $y_1 \in \{0,1\}$	$u_{6,1} = -1.4784x_1$ $- 2.8756y_1 - 4.7063$	$x_2 = -0.7986x_1$ $- 0.0491y_1 + 0.0568$ $y_2 = 1$

problem. The global optimum lies in $CR_{6,1}$ and the optimal decision variables are $x_1 = 10$, $y_1 = 0$, $x_2 = -7.9297$, $y_2 = 1$, $x_3 = -14.4202$, $y_3 = 0$ and $y_4 = 0$.

Table 3.8: Example 8: First-level problem solutions.

CR	Objectives	Decision variables
$CR_{1,1}$	$z_{1,1} = 41.2188, u_{1,1} = 8.5542,$ $v = -40.4303$	$x_1 = -7.2864, y_1 = 0, x_2 = 4.8932$ $y_2 = 1, x_3 = 9.1609, y_3 = 0, y_4 = 0$
$CR_{1,2}$	$z_{1,2} = -52.0008, u_{1,2} = 0.9185,$ $v = -16.6746$	$x_1 = 10, y_1 = 0, x_2 = -8.9996$ $y_2 = 1, x_3 = 4.9189, y_3 = 0, y_4 = 0$
$CR_{2,1}$	$z_{2,1} = 39.4173, u_{2,1} = 15.6712,$ $v = -36.3964$	$x_1 = -7.0258, y_1 = 0, x_2 = 4.8816$ $y_2 = 0, x_3 = 8.5270, y_3 = 1, y_4 = 0$
$CR_{2,2}$	$z_{2,2} = -52.3966, u_{2,2} = 8.1506,$ $v = -12.9990$	$x_1 = 10, y_1 = 0, x_2 = -8.8017$ $y_2 = 0, x_3 = 4.3489, y_3 = 1, y_4 = 0$
$CR_{3,1}$	$z_{3,1} = -52.1747, u_{3,1} = 0.5809$ $v = -15.4985$	$x_1 = 10, y_1 = 0, x_2 = -8.9126$ $y_2 = 1, x_3 = 4.6683, y_3 = 0, y_4 = 0$
$CR_{4,1}$	$z_{4,1} = -52.5680, u_{4,1} = -9.5001$ $v = 57.4323$	$x_1 = 10, y_1 = 0, x_2 = -8.7160$ $y_2 = 0, x_3 = -13.2161, y_3 = 1, y_4 = 0$
$CR_{5,1}$	$z_{5,1} = 32.2112, u_{5,1} = 4.1392$ $v = -25.2609$	$x_1 = -5.9831, y_1 = 0, x_2 = 4.8352$ $y_2 = 1, x_3 = -14.4202, y_3 = 0, y_4 = 0$
$CR_{5,2}$	$z_{5,2} = -53.9798, u_{5,2} = -2.9208$ $v = -3.2965$	$x_1 = 10, y_1 = 0, x_2 = -8.0101$ $y_2 = 1, x_3 = 2.0691, y_3 = 0, y_4 = 0$
$CR_{6,1}$	$z_{6,1} = -54.1407, u_{6,1} = -19.4906$ $v = 62.8215$	$x_1 = 10, y_1 = 0, x_2 = -7.9297$ $y_2 = 1, x_3 = -14.4202, y_3 = 0, y_4 = 0$

The computational performance of the algorithm for this example is presented in Table 3.9.

Table 3.9: Example 2 computational performance.

Problem	Problem type	Number of problems	CPU time (s)
Third Level	mp-MILP	1	1.159315
Second Level	mp-MILP	6	0.912661
First Level	MILP	9	0.152264
Comparison	ILP	1	0.070477

3.2 Mixed-integer quadratic trilevel optimization

The algorithm presented in the previous section can be extended to cover problems where the objective functions of each decision level can have quadratic terms. The general form of the class of problems we tackle here is presented below:

$$
\begin{aligned}
\min_{x_1, y_1} \quad & x_1^T Q_1 x_1 + c_1^T \omega + c_{c1} \\
\text{s.t.} \quad & A_1 x + E_1 y \leq b_1 \\
& \min_{x_2, y_2} \quad [x_1^T \; x_2^T] Q_2 [x_1^T \; x_2^T]^T + c_2^T \omega + c_{c2} \\
& \text{s.t.} \quad A_2 x + E_2 y \leq b_2 \\
& \qquad \min_{x_3, y_3} \quad \omega^T Q_3 \omega + c_3^T \omega + c_{c3} \\
& \qquad \text{s.t.} \quad A_3 x + E_3 y \leq b_3 \\
& x \in \mathbb{R}^n, \quad y \in \{0, 1\}^m \\
& x = [x_1^T \; x_2^T \; x_3^T]^T, \quad y = [y_1^T \; y_2^T \; y_3^T]^T, \quad \omega = [x_1^T \; y_1^T \; x_2^T \; y_2^T \; x_3^T \; y_3^T]^T
\end{aligned}
\tag{3.21}
$$

where ω is a vector of all decision variables of all decision levels, x_i are continuous and y_i binary decision variables of optimization level i, $Q_i > 0, c_i$ and c_{ci} are constant coefficient matrices in the objective function of optimization level i, A_i, E_i are constant coefficient matrices multiplying decision variables of level i in the constraint set, and b is a constant value vector.

This algorithm follows the same methodology as the T-MILP algorithm presented in the previous section. It is based on a recently developed multiparametric mixed-integer quadratic programming algorithm [199]. The algorithm is summarized in 8 steps in Table 3.10, and then illustrated through a numerical example.

Table 3.10: Multiparametric based algorithm for the solution of T-MILP problems.

Step 1	Establish integer and continuous variable bounds, and transform the T-MIQP into a binary T-MIQP
Step 2	Recast the third-level problem as a mp-MIQP, in which the optimization variables of the second- and first-level problems are considered as parameters
Step 3	Solve the resulting mp-MIQP problem to obtain the optimal solution of the lower level problem as explicit functions of the second- and first-level decision variables
Step 4	Substitute each multiparametric solution into the second-level problem to formulate k mp-MIQP problems, considering the first-level decision variables as parameters
Step 5	Solve the resulting k mp-MIQP problems to obtain the optimal solution of the second-level problem as explicit functions of the first-level decision variables
Step 6	Substitute each multiparametric solution into the first-level problem to formulate single-level MIQP problems
Step 7	Solve all single-level problems using CPLEX® MIQP solver
Step 8	Solve the comparison optimization problem (3.22) to select the exact and global optimum solution

Remark 3.1. Clusters of solutions (overlapping critical regions) might appear for some critical regions due to the existence of binary variables. The exact solution would re-

quire extra computational time for a comparison procedure. For the case of mp-MIQP problems, this could also result into nonconvex regions that would be difficult to handle in the following steps. A comparison procedure at the end of this algorithm is computationally much more efficient.

The comparison optimization problem for T-MIQP problems is presented below:

$$
\begin{aligned}
z^* = \min_{\alpha, y} \quad & \alpha \\
\text{s.t.} \quad & \alpha = \sum_{i,j} y_{i,j} z_{i,j} && [c_1] \\
& \sum_{i,j} y_{i,j} = 1 && [c_2] \\
& y_{i,j} u_{i,j} \leq y_{i,j} u_{p,q} \quad \forall i,j,p \neq i,q && [c_3] \\
& y_{i,j} v_i \leq y_{i,j} v_r \quad \forall i,j,r \neq i && [c_4] \\
& y_{i,j} \in \{0,1\}
\end{aligned}
\tag{3.22}
$$

where z^* is the exact global optimum of the original trilevel MIQP programming problem, $y_{i,j}$ are binary variables corresponding to each $CR_{i,j}$, $z_{i,j}$ are the objective function values obtained when solving problems in Step 7, u_i are the objective function values obtained when solving problems in Step 5, and v_i are the objective function values obtained when solving problems in Step 3. Constraint $[c_2]$ makes sure that only one $y_{i,j}$ is equal to 1 and the rest are equal to zero (i. e., only one solution is selected), constraint $[c_3]$ makes sure that the selected solution results to an optimal solution in the second level, and constraint $[c_4]$ makes sure that the selected solution results to an optimal solution in the third level. When $y_{i,j} = 1$, then the optimal solution lies in $CR_{i,j}$ and CR_i.

3.2.1 Numerical example

3.2.1.1 Example 9: Trilevel mixed-integer quadratic programming problem

Consider the following trilevel T-MIQP problem (3.23) of Type 3:

$$
\begin{aligned}
\min_{x_1, y_1} \quad & z = 5x_1^2 + 6x_2^2 + 3y_1 + 3y_2 - 3x_3 \\
\text{s.t.} \quad \min_{x_2, y_2} \quad & u = 4x_1^2 + 6y_1 - 2x_2 + 10y_2 - x_3 + 5y_3 \\
\text{s.t.} \quad \min_{x_3, y_3, y_4} \quad & v = 4x_3^2 + y_3^2 + 5y_4^2 + x_2 y_3 + x_2 y4 - 10x_3 - 15y_3 - 16y_4 \\
\text{s.t.} \quad & 6.4x_1 + 7.2x_2 + 2.5x_3 \leq 11.5 \\
& -8x_1 - 4.9x_2 - 3.2x_3 \leq 5 \\
& 3.3x_1 + 4.1x_2 + 0.02x_3 + 0.2y_1 + 0.8y_2 + 4y_3 + 4.5y_4 \leq 1 \\
& y_1 + y_2 + y_3 + y_4 \geq 1 \\
& -10 \leq x_1, x_2 \leq 10 \\
x_1, x_2, x_3 \in \mathbb{R}^3, \quad & y_1, y_2, y_3, y_4 \in \{0,1\}^4
\end{aligned}
\tag{3.23}
$$

The multiparametric solution of the mp-MIQP third-level problem consists of 14 critical regions, with a subset of them presented in Table 3.11.

Table 3.11: Example 9: A subset of the multiparametric solution of the third-level problem.

CR	Definition	3rd Level Obj.	3rd Level Var.
CR_1	$0.6236x_1 + 0.7759x_2$ $+ 0.0960y_2 \leq 0.1743$ $-0.6644x_1 - 0.7474x_2 \leq -0.2206$ $-y_1 - y_2 \leq -1$ $x_1 \leq 10\ y_1, y_2 \in \{0, 1\}$	$v_1 = 13.1072x_1{}^2$ $+ 16.5888x_2{}^2$ $+ 29.4912x_1x_2$ $- 8.7040x_1$ $- 9.7920x_2$ $- 26.6800$	$x_3 = -2.56x_1$ $\quad - 2.88x_2 + 4.6$ $y_3 = 0$ $y_4 = 0$
CR_6	$0.62119x_1 + 0.7778x_2$ $+ 0.0956y_2 \leq -0.5674$ $-0.6242x_1 - 0.7755x_2$ $+ -0.0946y_2 \leq 0.5816$ $y_1 + y_2 \leq 1$ $x_1 \leq 10$ $y_1, y_2 \in \{0, 1\}$	$v_6 = 54450x_1{}^2$ $+ 84050x_2{}^2$ $+ 1250y_2{}^2$ $+ 16.5888x_2{}^2$ $+ 135300x_1x_2$ $+ 16500x_1y_2$ $+ 20500x_2y_2$ $+ 101475x_1$ $+ 126076x_2$ $+ 15375y_2 + 15375$	$x_3 = -165x_1$ $\quad - 205x_2 - 25y_2$ $\quad - 150$ $y_3 = 1$ $y_4 = 0$

The multiparametric solution of the 14 mp-MIQP second-level problems consists of 22 critical regions. A subset of them is presented in Table 3.12.

Table 3.12: Example 9: A subset of the multiparametric solution of second-level problems.

CR	Definition	2nd level objective	2nd level var.
$CR_{1,1}$	$2.2792 \leq x_1 \leq 10$ $y_1 \in \{0, 1\}$	$u_{1,1} = 4x_1{}^2 + 1.7778x_1 + 6y_1 + 3.6597$	$x_2 = -0.8889x_1 + 0.2951$ $y_2 = 1$
$CR_{6,1}$	$-3.2852 \leq x_1 \leq 10$ $y_1 \in \{0, 1\}$	$u_{6,1} = 4x_1{}^2 + 1.6098x_1 + 6y_1 + 2.75$	$x_2 = -0.8049x_1 - 0.75$ $y_2 = 0$

The 22 single-level MIQP problems formulated are solved using CPLEX®mixed-integer quadratic programming solver. A subset of the resulting local solutions are presented in Table 3.13.

The exact global optimum is found to be lying in $CR_{6,1}$ with optimal decisions $x_1 = -0.4076$, $y_1 = 0$, $x_2 = -0.4220y_2 = 0$, $x_3 = 3.7500$, $y_3 = 1$ and $y_4 = 0$.

The computational performance of the algorithm for example 9 is presented in Table 3.14.

Table 3.13: Example 9: A subset of the first-level problem solutions.

CR	Objectives	Decision variables
$CR_{1,1}$	$z_{1,1} = 35.6995$, $u_{1,1} = 30.4913$, $v = 0$	$x_1 = 2.2792$, $y_1 = 0$, $x_2 = -1.7308$ $y_2 = 1$, $x_3 = 3.7500$, $y_3 = 0$, $y_4 = 0$
$CR_{6,1}$	$z_{6,1} = -9.3512$, $u_{6,1} = 2.7583$, $v = -43.0470$	$x_1 = -0.4076$, $y_1 = 0$, $x_2 = -0.4220$ $y_2 = 0$, $x_3 = 3.7500$, $y_3 = 1$, $y_4 = 0$

Table 3.14: Example 9 computational performance.

Problem	CPU time (s)
Third-level mp-MIQP	4.427805
Second-level mp-MIQPs (14)	1.743407
First-level MIQPs (22)	0.129060
Comparison	0.012595

3.3 Extensions to more general multilevel optimization problems

3.3.1 Multilevel mixed-integer nonlinear optimization

The algorithms presented in Chapter 2 and Chapter 3, Section 3.1 can be extended to cover problems with more than three optimization levels. Furthermore, the objective function of the first-level problem can be a general mixed-integer nonlinear function of all the decision variables of the multilevel problem.

The general formulation of the class of problems considered in this section is presented below:

$$
\begin{aligned}
&\min_{x_1, y_1} \quad Z_{L1} = F_1(\omega) \\
&\text{s.t.} \quad A_1 x + E_1 y \le b_1 \\
&\qquad \min_{x_2, y_2} \quad Z_{L2} = [x_1^T x_2^T] Q_2 [x_1^T x_2^T]^T + c_2^T \omega + c_{c2} \\
&\qquad \text{s.t.} \quad A_2 x + E_2 y \le b_2 \\
&\qquad\qquad \ddots \\
&\qquad\qquad \min_{x_n, y_n} \quad Z_{Ln} = \omega^T Q_n \omega + c_n^T \omega + c_{cn} \\
&\qquad\qquad \text{s.t.} \quad A_n x + E_n y \le b_n \\
&\quad x = \begin{bmatrix} x_1^T & x_2^T & \cdots & x_n^T \end{bmatrix}^T, \quad y = \begin{bmatrix} y_1^T & y_2^T & \cdots & y_n^T \end{bmatrix}^T \\
&\quad \omega = \begin{bmatrix} x_1^T & y_1^T & x_2^T & y_2^T & \cdots & x_n^T y_n^T \end{bmatrix}^T \\
&\quad x \in \mathbb{R}^m, \quad y \in \{0,1\}^p
\end{aligned}
\tag{3.24}
$$

where ω is a vector of all decision variables of all optimization levels, x_i are continuous and y_i binary decision variables of optimization level i, x is a vector of all continuous variables of all optimization levels, y is a vector of all binary variables of all optimization levels, F_1 is a nonlinear, nonconvex function of all decision variables, $Q_i > 0, c_i$ and c_{ci} are constant coefficient matrices in the objective function of optimization level i, A_i are constant coefficient matrices multiplying continuous decision variables, x, in level i, E_i are constant coefficient matrices multiplying binary decision variables, y, in level i in the constraint set and b_i are constant value vectors in the constraint set of level i.

This algorithm follows the same methodology as the T-MILP and T-MIQP algorithms presented in the previous section of this chapter. When the objective function of the first-level problem is nonlinear, CPLEX® is no longer used. Instead, solvers capable of providing the global optimum of mixed-integer nonlinear problems, such as ANTIGONE® [246] and BARON® [247], are used.

The comparison optimization problem for M-MINLP problems is presented below:

$$z^* = \min_{\alpha,\Gamma} \quad \alpha$$

$$\text{s. t.} \quad \alpha = \sum_{i_1,i_2,\dots,i_n} \Gamma_{i_1,i_2,\dots i_n} Z_{L1,i_1,i_2,\dots,i_n} \qquad [c_1]$$

$$\sum_{i_1,i_2,\dots i_n} \Gamma_{i_1,i_2,\dots i_n} = 1 \qquad [c_2] \qquad (3.25)$$

$$\Gamma_{i_1,i_2,\dots i_n} Z_{Lm,i_1,i_2,\dots i_n} \le \Gamma_{i_1,i_2,\dots i_n} Z_{Lm,j_1,j_2\dots j_n} \quad \forall i,j,m \le 2 \quad [c_3]$$

$$\Gamma_{i_1,i_2,\dots i_n} \in \{0,1\} \quad \forall i$$

where z^* is the exact global optimum of the original multilevel mixed-integer nonlinear programming problem, $\Gamma_{i_1,i_2,\dots i_n}$ are binary variables corresponding to each critical region i created at every optimization level 1 to n, and $Z_{Lk,i_1,i_2,\dots i_n}$ are the k level objective function values at the optimal decision of corresponding to $Z_{L1,i_1,i_2,\dots i_n}$. Constraint $[c_2]$ makes sure that only one $\Gamma_{i_1,i_2,\dots i_n}$ is equal to 1 and the rest are equal to zero (i. e., only one solution is selected); constraint $[c_3]$ makes sure that the selected solution results to an optimal solution in the rest of the decision levels. When $\Gamma_{i_1,i_2,\dots i_n}$, then the optimal solution lies in $CR_{i_1,i_2,\dots i_n}$.

The proposed algorithm is summarized in Table 3.15, and then illustrated through a numerical example.

3.3.1.1 Example 10: Multilevel mixed-integer nonlinear programming problem

Consider the following four-level mixed-integer nonlinear programming problem (3.26) of Type 3. The first-level problem is a MINLP problem, the second- and the fourth-level problems are MILP problems, and finally the third-level problem is a MIQP problem. The M-MINLP problem has four continuous and seven binary vari-

Table 3.15: Multiparametric based algorithm for the solution of multilevel mixed-integer optimization problems.

Step 1	Establish integer and continuous variable bounds, and transform the M-MIQP into a binary M-MIQP. Set the number of optimization levels to n and $k = n - 1$
Step 2	Recast the last level problem as a mp-MIQP, in which the optimization variables of all problems above the last level problem are considered as parameters
Step 3	Solve the resulting mp-MIQP problem to obtain the optimal solution of the lower-level problem as explicit functions of all other decision variables
Step 4	Substitute each multiparametric solution into the kth level problem to formulate mp-MIQP problems, considering the optimization variables of all problems above level k as parameters
Step 5	Solve the resulting mp-MIQP problems to obtain the optimal solution of the kth level problem as explicit functions of the upper levels decision variables
Step 6	Set $k = k - 1$ and go to Step 4. Continue to Step 7 when $k = 1$
Step 7	Substitute each multiparametric solution into the first-level problem to formulate single-level MINLP problems
Step 8	Solve all single-level problems using CPLEX® MILP or MIQP solvers, ANTIGONE®or BARON®solvers
Step 9	Solve the comparison optimization problem (3.25) to select the global optimum solution

ables:

$$
\begin{aligned}
\min_{x_1, y_1} \quad & Z_{L1} = 4x_1^{\,3} - 4x_1^{\,4} + x_1^{\,2}x_3 + x_2x_4 - 7x_1 - 7y_{1,3} - 2x_2 - y_{1,2} \\
\text{s.t.} \quad \min_{x_2, y_2} \quad & Z_{L2} = 5x_2 + 6x_3 + 2x_4 + 3y_2 + 3y_3 \\
\text{s.t.} \quad \min_{x_3, y_3} \quad & Z_{L3} = 4x_2^{\,2} + 4y_3 - 2x_3 - x_4 + 5y_{4,1} \\
\text{s.t.} \quad \min_{x_4, y_4} \quad & Z_{L4} = 2x_2 + y_2 + 2x_3 + y_3 - 4x_4 + 2y_{4,1} \\
\text{s.t.} \quad & x_1 + 6.4x_2 + 7.2x_3 + 2.5x_4 \le 11.5 \\
& 0.1x_1 + 3.3x_2 + 4.1x_3 + 0.02x_4 + 3y_{1,1} \le 1 \\
& 0.2y_2 + 0.8y_3 + 4y_{4,1} + 4.5y_{4,2} - 4y_{1,1} \le 2 \\
& y_2 + y_3 + y_{4,1} + y_{4,2} \ge 1 \\
& y_{1,1} + y_{1,2} + y_{1,3} \le 2 \\
& -10 \le x_1, x_2, x_3, x_4 \le 10 \\
& x_1, x_2, x_3, x_4 \in \mathbb{R}^4, \quad y_{1,1}, y_{1,2}, y_{1,3}, y_2, y_3, y_{4,1}, y_{4,2} \in \{0,1\}^7
\end{aligned}
\tag{3.26}
$$

The multiparametric solution of the fourth-level mp-MILP problem consists of 8 critical regions, with a subset of them presented in Table 3.16.

Table 3.16: Example 10: A subset of the multiparametric solution of the fourth-level problem.

CR	Definition	4th level obj.	4th level var.
CR_1	$-y_2 - y_3 \leq -1$ $y_{1,1} + y_{1,2} + y_{1,3} \leq 2$ $-0.0013x_1 + 0.5007y_{1,1}$ $\quad + 0.5423x_2 + 0.6747x_3 \leq 0.1516$ $-0.0245x_1 + 0.9791y_{1,1}$ $\quad - 0.0490y_2 - 0.1958y_3 \leq 0.4896$ $-x_1 \leq 10$ $-10 \leq x_2 \leq 10$ $-x_3 \leq 10$ $y_{1,1}, y_{1,2}, y_{1,3}, y_2, y_3 \in \{0, 1\}$	$Z_{L4,1} = 1.6x_1$ $\quad + 12.24x_2$ $\quad + y_2 + 13.52x_3$ $\quad + y_3 - 18.4$	$x_4 = -0.4x_1$ $\quad - 2.56x_2$ $\quad - 2.88x_3 + 4.6$ $y_{4,1} = 0$ $y_{4,2} = 0$
CR_8	$y_2 + y_3 \leq 1$ $y_{1,1} + y_{1,2} + y_{1,3} \leq 2$ $0.0013x_1 - 0.5007y_{1,1}$ $\quad - 0.5423x_2 - 0.6747x_3 \leq -0.1516$ $0.0245x_1 - 0.9791y_{1,1}$ $\quad + 0.0490y_2 + 0.1958y_3 \leq -0.4896$ $-0.0245x_1 + 0.9791y_{1,1}$ $\quad - 0.0490y_2 + 0.1958y_3 \leq 0.6119$ $-10 \leq x_1 \leq 10$ $-10 \leq x_2 \leq 10$ $x_3 \leq 10$ $y_{1,1}, y_{1,2}, y_{1,3}, y_2, y_3 \in \{0, 1\}$	$Z_{L4,8} = 600y_{1,1}$ $\quad + 662x_2$ $\quad + y_2 + 822x_3$ $\quad + y_3 - 198$	$x_4 = -150y_{1,1}$ $\quad - 165x_2$ $\quad - 205x_3 + 50$ $y_{4,1} = 1$ $y_{4,2} = 0$

The multiparametric solution of the eight third-level mp-MIQP problems consists of 8 critical regions, with a subset of them presented in Table 3.17.

The multiparametric solution of the eight second-level mp-MILP problems consists of 17 critical regions, with a subset of them presented in Table 3.18.

The 17 single-level problems corresponding to the first level were solved using ANTIGONE® solver through GAMS®. The solution for a subset of the problems is given in Table 3.19.

After the comparison procedure, the global optimum solution lies in $CR_{1,1,1}$ at the point presented in Table 3.19.

3.3.2 Mixed-integer multifollower optimization

Optimization problems involving a leader with multiple followers are referred to as bi-level multifollower programming problems. Procedures for the solution of mixed-integer multifollower optimization problems include very limited heuristic approaches, such as Sinha et al. [154] that do not guarantee global optimality.

Table 3.17: Example 10: A subset of the multiparametric solutions of the third-level problem.

CR	Definition	4th level obj.	4th level var.
$CR_{1,1}$	$y_{1,1} + y_{1,2} + y_{1,3} \le 2$ $-0.0250x_1 + 0.9984y_{1,1}$ $\quad - 0.0499y_2 \le 0.6989$ $-10 \le x_1 \le 10$ $-10 \le x_2 \le 10$ $y_{1,1}, y_{1,2}, y_{1,3}, y_2 \in \{0, 1\}$	$Z_{L3,1,1} = 0.4x_1$ $\quad + 2.56x_2 - 11.4$	$x_3 = -10$ $y_3 = 1$
$CR_{8,1}$	$y_{1,1} + y_{1,2} + y_{1,3} \le 2$ $-0.0250x_1 + 0.9984y_{1,1}$ $\quad - 0.0499y_2 \le 0.6240$ $0.0250x_1 - 0.9984y_{1,1}$ $\quad + 0.0499y_2 \le -0.4992$ $-10 \le x_1 \le 10$ $-10 \le x_2 \le 10$ $y_{1,1}, y_{1,2}, y_{1,3}, y_2 \in \{0, 1\}$	$Z_{L3,8,1} = 0.417x_1$ $\quad - 0.6531y_{1,1}$ $\quad + 1.8528x_2 + 0.5977$	$x_3 = 0.0020x_1$ $\quad - 0.7421y_{1,1}$ $\quad - 0.8037x_2 + 0.2246$ $y_3 = 0$

Table 3.18: Example 10: A subset of the multiparametric solutions of the second-level problem.

CR	Definition	4th level obj.	4th level var.
$CR_{1,1,1}$	$y_{1,1} + y_{1,2} + y_{1,3} \le 2$ $-0.0250x_1 + 0.9984y_{1,1} \le 0.6989$ $-10 \le x_1 \le 10$ $y_{1,1}, y_{1,2}, y_{1,3}, y_2 \in \{0, 1\}$	$Z_{L2,1,1,1} = -0.8x_1 + 28.6$	$x_2 = 10$ $y_2 = 0$
$CR_{1,1,2}$	$y_{1,1} + y_{1,2} + y_{1,3} \le 2$ $-0.0250x_1 + 0.9984y_{1,1} \le 0.7498$ $0.0250x_1 - 0.9984y_{1,1} \le -0.6998$ $-10 \le x_1 \le 10$ $y_{1,1} = 0$ $y_{1,2}, y_{1,3}, y_2 \in \{0, 1\}$	$Z_{L2,1,1,2} = -0.8x_1 + 31.6$	$x_2 = 10$ $y_2 = 1$

Table 3.19: Example 10: A subset of the first-level problem solutions.

CR	Objectives	Decision variables
$CR_{1,1,1}$	$Z_{L1,1,1,1} = -2052.270$ $Z_{L2,1,1,1} = 31.824$ $Z_{L3,1,1,1} = 414.588$ $Z_{L4,1,1,1} = -36.648$	$x_1 = -4.030, y_{1,1} = 0, y_{1,2} = 1, y_{1,3} = 1$ $x_2 = 10, y_2 = 0$ $x_3 = -10, y_3 = 1$ $x_4 = 9.412, y_{4,1} = 0, y_{4,2} = 0$
$CR_{1,1,2}$	$Z_{L1,1,1,2} = 45633$ $Z_{L2,1,1,2} = 23.6$ $Z_{L3,1,1,2} = 420.2$ $Z_{L4,1,1,2} = -13.2$	$x_1 = 10, y_{1,1} = 1, y_{1,2} = 0, y_{1,3} = 1$ $x_2 = 10, y_2 = 1$ $x_3 = -10, y_3 = 1$ $x_4 = 3.8, y_{4,1} = 0, y_{4,2} = 0$

The proposed global solution algorithm will be introduced through the general form of the bilevel multifollower mixed-integer linear programming (BMF-MILP) problem (3.27), and then illustrated through a case study on a planning and scheduling integration example in Section 4.4 and a design and scheduling integration example in Section 4.5 of this book.

$$
\begin{aligned}
&\min_{x_1, y_1} \quad Q_1 x + H_1 y + c_{c1} \\
&\text{s.t.} \quad A_1 x + E_1 y \le b_1 \\
&\qquad \min_{x_{2,a}, y_{2,a}} \quad Q_{2,a} x + H_{2,a} y + c_{c2,a} \\
&\qquad \text{s.t.} \quad A_{2,a} x + E_{2,a} y \le b_{2,a} \\
&\qquad \min_{x_{2,b}, y_{2,b}} \quad Q_{2,b} x + H_{2,b} y + c_{c2,b} \\
&\qquad \text{s.t.} \quad A_{2,b} x + E_{2,b} y \le b_{2,b} \\
&\qquad \qquad \vdots \\
&\qquad \min_{x_{2,n}, y_{2,n}} \quad Q_{2,n} x + H_{2,n} y + c_{c2,n} \\
&\qquad \text{s.t.} \quad A_{2,n} x + E_{2,n} y \le b_{2,n} \\
&\qquad x = \begin{bmatrix} x_1^T & x_{2,a}^T & x_{2,b}^T & \cdots & x_{2,n}^T \end{bmatrix}^T, \quad x \in \mathbb{R}^m \\
&\qquad y = \begin{bmatrix} y_1^T & y_{2,a}^T & y_{2,b}^T & \cdots & y_{2,n}^T \end{bmatrix}^T, \quad y \in \{0,1\}^p
\end{aligned}
\tag{3.27}
$$

It is assumed that upper level optimization variables that appear in the lower level problems, and lower level integer variables are bounded.

As a first step, the lower level problems are transformed into multiparametric mixed-integer problems (3.28). For each transformed lower level problem, the optimization variables of the leader problem along with any decision variables of the rest of the lower level problems that appear in the transformed problem are considered as parameters:

$$
\begin{aligned}
&\min_{x_{2,a}, y_{2,a}} \quad Q_{2,a} x + H_{2,a} y + c_{c2,a} \\
&\text{s.t.} \quad A_{2,a} x + E_{2,a} y \le b_{2,a} \\
&\qquad x^L \le x \le x^U \\
&\min_{x_{2,b}, y_{2,b}} \quad Q_{2,b} x + H_{2,b} y + c_{c2,b} \\
&\text{s.t.} \quad A_{2,b} x + E_{2,b} y \le b_{2,b} \\
&\qquad x^L \le x \le x^U \\
&\qquad \qquad \vdots \\
&\min_{x_{2,n}, y_{2,n}} \quad Q_{2,n} x + H_{2,n} y + c_{c2,n} \\
&\text{s.t.} \quad A_{2,n} x + E_{2,n} y \le b_{2,n} \\
&\qquad x^L \le x \le x^U
\end{aligned}
\tag{3.28}
$$

The solution of the multiparamertric problems (3.28), using mp-MILP or mp-MIQP algorithms through POP® toolbox [242], provides the complete profile of optimal solutions of the lower level problems as explicit affine functions of the variables of the upper level problem and other lower level problems within corresponding critical regions (3.29). The problems (3.28) are interdependent of each other, therefore, parallel programming can be used to solve them simultaneously:

$$[x_{2,n}y_{2,n}] = \begin{cases} \xi_{2,n}^1 = p_{2,n}^1 + q_{2,n}^1 x, \psi_{2,n}^1 & \text{if } H_{2,n}^1 x \le h_{2,n}^1 \\ \xi_{2,n}^2 = p_{2,n}^2 + q_{2,n}^2 x, \psi_{2,n}^2 & \text{if } H_{2,n}^2 x \le h_{2,n}^2 \\ \quad \vdots & \quad \vdots \\ \xi_{2,n}^k = p_{2,n}^k + q_{2,n}^k x, \psi_{2,n}^k & \text{if } H_{2,n}^k x \le h_{2,n}^k \end{cases} \tag{3.29}$$

where $\xi_{2,n}^i$ is the lower level $2, n$ objective, $H_{2,n}^i x \le h_{2,n}^i$ is referred to as critical region, CR^i, and k denotes the number of computed critical regions.

As a next step, the solutions derived in the previous step are used to formulate a set of reformulations of the original bilevel multifollower problem. Each reformulation is formed by using a different combination of critical regions from the solutions of the follower's problems (3.29) to achieve a Nash equilibrium. Considering only the upper level problem, the follower's decision variables are substituted by the affine functions derived. The critical region definitions are added to the upper level problem as a new set of constraints along with constraints for Nash equilibrium, forming single-level deterministic problems for every different combination of critical regions. In the case of quadratic objective functions, the single-level deterministic problems created in this step can contain nonlinear constraints as critical region definitions can be nonlinear. This makes the single-level problems more challenging to be solved.

Single-level problems with the formulation of (3.30) and all possible combinations of $i, j, \ldots k$ are created:

$$\begin{aligned} \min_{x_1, y_1} \quad & Q_1 x + H_1 y + c_{c1} \\ \text{s.t.} \quad & A_1 x + E_1 y \le b_1 \\ & H_{2,a}^i x \le h_{2,a}^i \\ & x_{2,a} = p_{2,a}^i + q_{2,a}^i x \\ & y_{2,a} = \psi_{2,a}^i \\ & H_{2,b}^j x \le h_{2,b}^j \\ & x_{2,b} = p_{2,b}^j + q_{2,b}^j x \\ & y_{2,b} = \psi_{2,b}^j \\ & \quad \vdots \\ & H_{2,n}^k x \le h_{2,n}^k \\ & x_{2,n} = p_{2,n}^k + q_{2,n}^n x \\ & y_{2,n} = \psi_{2,n}^k \end{aligned} \tag{3.30}$$

All single-level MI(N)LP problems formed are solved using appropriate solvers, such as CPLEX®or ANTIGONE®solvers. Because those problems are independent of each other, it is possible to use parallel programming to solve them simultaneously.

The solutions of the above single-level MI(N)LP problems correspond to different local optimal solutions of the original BMF-MILP. The final step of the algorithm is to compare all the local solutions to obtain the minimum z that would correspond to the exact and global optimum, z^*, of the original bilevel multifollower problem.

The proposed algorithm is also summarized in Table 3.20.

Table 3.20: Multiparametric based algorithm for the solution of BMF-MILP problems.

Step 1	Establish integer and continuous variable bounds
Step 2	Recast the lower level problems as mp-MILPs, in which the optimization variables of the upper level problem along with the optimization variables of other lower level problems are considered as parameters
Step 3	Solve the resulting mp-MILP problems to obtain the optimal solution of the lower level problems as explicit functions of the upper level variables and other lower level variables
Step 4	Substitute each multiparametric solution into the upper level problem and add Nash equilibrium constraints to formulate k single-level MILP problems
Step 5	Solve all k single-level problems and compare their solutions to select the exact and global optimum

3.3.3 Conclusion

This chapter discussed algorithms for the exact global solution of more general classes of mixed-integer multilevel optimization problems with integer and continuous variables in all optimization levels. The algorithms presented utilize multi-parametric programming to solve the lower level optimization problems as a function of all of the upper levels problem variables.

The next chapter will introduce different application case studies, including classical bilevel problems such as the integration of production and distribution planning, and other novel applications such as a hierarchical economic model predictive controller and a class of robust optimization. The formulation and solution of these problems will be presented in detail.

4 Application areas

In order to highlight the applicability of the algorithms presented in the two previous chapters of this book, this chapter introduces six different application areas for all the different classes of problems explored.

For all case studies, the problem formulation and solution approach is explained in detail.

4.1 B-MILP and mp-B-MILP: distribution and production planning

We are considering a case where the Production-Distribution planning (PD) problem can be expressed as a hierarchical decision problem, involving two different decision makers corresponding to each company that owns the Production plants and Distribution centers. Assuming one company owns the Production plants and another the Distribution centers, the resulting problem is a two level hierarchical decision problem. The first level is responsible for optimizing the Distribution centers overall costs and is influenced by the second level that is responsible for optimizing the Production plants overall costs.

When considering the PD problem, decisions taken at both decision levels can involve both continuous (e. g., Production rates, Distribution rates) or discrete (e. g., choice of Production plant, choice of Distribution center, active routes) variables. Therefore, the integrated PD problem results into a bilevel mixed-integer programming problem.

4.1.1 Uncertainty in supply chain systems

In the existing unstable business environment, with constantly changing market conditions and customer needs and expectations, it is of high importance to consider the effect of uncertainties in the integrated PD problem. Sources of uncertainty in supply chain planning may include variations in processing rates, canceled or rushed orders, equipment failure, raw material, final product or utility price fluctuations and demand variations [35].

In this study, we will consider one of the key sources of uncertainty in the PD system, the product demand. Failure to consider this in PD planning could lead to either unsatisfied customer demand and loss of market share, or excessively high inventory holding costs [36]. To address this, a number of publications have been devoted in studying supply chain planning under demand uncertainty [26, 28, 33, 248, 249].

https://doi.org/10.1515/9783110760316-004

4.1.2 Bilevel programming model for a supply chain planning problem

We will consider a supply chain system operated by two different companies (Figure 4.1), one owning two Production plants, P1 and P2 that produce one product that is distributed by the other company owning two Distribution centers, D1 and D2, to three different costumers, C1, C2 and C3, located at various locations.

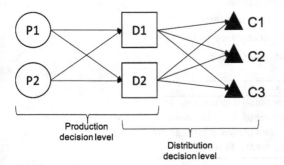

Figure 4.1: Schematic representation of the Production–Distribution planning problem case study with all the possible routes connecting Production plants with Distribution centers and customers. P1 and P2 are the two Production plants, D1 and D2 are the two Distribution centers and C1, C2 and C3 are the three customers.

The Distribution company decides the allocation of costumers to each Distribution center, the existence or not of the routes connecting the centers to the costumers and, therefore, the required inventory level in each center. Its objective is to minimize the cost of distributing and acquiring the items from the processing plants.

The Production company, after receiving the required inventory level for each Distribution center, decides which processing plant will take the orders, by minimizing Production costs. The Production constraints ensure that the allocation of orders to the plants satisfies capacity and Production target.

Therefore, we can consider the Distribution decision level to be the upper level (leader), deciding on the required inventory level of each Distribution center, while not knowing which plants will be supplying the centers to achieve that inventory level, but knowing that the Production decision level (lower level, follower) will take that decision by minimizing its own cost.

Two cases will be considered. In Case I, the demands of all three customers will be considered to be known and constant, whereas in Case II this demand will be considered as a bounded uncertainty.

4.1.2.1 Case I: Constant demand (B-MILP)

A simple Production–Distribution bilevel mixed-integer model for problem Case I is formulated below (4.1), using the notation given in Table 4.1.

Table 4.1: Notation.

Sets and index	
P, p	Set of plants, plant index
D, d	Set of Distribution centers, centers index
C, c	Set of customers, customers index
Constants	
A_p	Maximum capacity of plant p
c^1_{dc}	Cost of route connecting center d to costumer c
c^2_{pd}	Cost of getting products from plant p to center d
c^3_{pd}	Cost of manufacturing a product at plant p for center d
c^4_{pd}	Cost of route connecting plant p to center d
b_c	Customer c demand
Decision variables	
s_{dc}	Amount of product send from center d to customer c
y_{dc}	Existence of route connecting center d to customer c
x_{pd}	Amount of product manufactured in plant p for center d
z_{pd}	Existence of route connecting plant p to center d

Expressions (a) to (d) correspond to the leader (upper level) Distribution problem and (e) to (i) to the followers (lower level) Production problem. Expressions (a) and (e) are the leader's and follower's objective functions, respectively, and reflect the cost minimization of each company.

Constraint (b) ensures that the Distribution centers supply enough product to meet the demand of each costumer. Constraint (c) is a big-M constraint formulation expression, limiting the value of the amount of product send from each Distribution center to the costumers to less than 35 when the route connecting the two is active, but forcing it to zero, along with constraint (d), when the route is not active.

$$
\begin{aligned}
& \min_{y_{dc}, s_{dc}} \quad \sum_{d \in D} \sum_{c \in C} c^1_{dc} y_{dc} + \sum_{p \in P} \sum_{d \in D} c^2_{pd} x_{pd} && \text{(a)} \\
& \text{s.t.} \quad \sum_{d \in D} s_{dc} \geq b_c, && \forall c \in C && \text{(b)} \\
& \qquad s_{dc} \leq 35 y_{dc}, && \forall c \in C, \quad \forall d \in D && \text{(c)} \\
& \qquad s_{dc} \geq 0, && \forall c \in C, \quad \forall d \in D && \text{(d)} \\
& \min_{x_{pd}, z_{pd}} \quad \sum_{p \in P} \sum_{d \in d} c^3_{pd} x_{pd} + c^4_{pd} z_{pd} && \text{(e)} \\
& \text{s.t.} \quad \sum_{d \in D} x_{pd} \leq A_p && \forall p \in P && \text{(f)} \\
& \qquad \sum_{p \in P} x_{pd} \geq \sum_{c \in C} s_{dc} && \forall d \in D && \text{(g)} \\
& \qquad x_{pd} \leq 100 z_{pd} && \forall d \in D, \quad \forall p \in P && \text{(h)} \\
& \qquad x_{pd} \geq 0 && \forall d \in D, \quad \forall p \in P && \text{(i)} \\
& \quad s_{dc} \in \mathbb{R}, \quad y_{dc} \in \{0, 1\}, \quad x_{pd} \in \mathbb{R}, \quad z_{pd} \in \{0, 1\}
\end{aligned}
$$

(4.1)

Constraint (f) limits the Production of each plant to their maximum capacity, and constraint (i) ensures that the Production is positive. Constraint (g) ensures that the plants will produce enough to satisfy the demand of the Distribution centers. Constraint (h) is similar to (c), a typical big-M formulation expression, limiting the value of product produced for each Distribution center to less than 100 when the route connecting the Distribution centers to the plants is active, but forcing it to zero, along with constraint (i), when the route is not active.

The constants used in example Case I are presented in Table 4.2.

Table 4.2: Constants for PD Case study.

A_p		$p = 1$	$p = 2$	
		135	100	
c_{dc}^1		$c = 1$	$c = 2$	$c = 3$
	$d = 1$	75	60	50
	$d = 2$	80	30	65
c_{pd}^2		$d = 1$	$d = 2$	
	$p = 1$	21	30	
	$p = 2$	26	25	
c_{pd}^3		$d = 1$	$d = 2$	
	$p = 1$	20	25	
	$p = 2$	20	25	
c_{pd}^4		$d = 1$	$d = 2$	
	$p = 1$	100	80	
	$p = 2$	110	70	
b_c		$c = 1$	$c = 2$	$c = 3$
		55	65	15

4.1.2.2 Case 2: Demand uncertainty (mp-B-MILP)

This case considers demand to be unknown but bounded. In order to express this, equation (4.2) is added to the model presented before. b_c is removed from the set of constants, and is now treated as a parameter of the optimization problem:

$$0 \le b_c \le 70, \quad \forall c \in C \tag{4.2}$$

4.1.3 Solving the supply chain planning problem

4.1.3.1 Case I: Constant demand (B-MILP)

The first and second steps in the algorithm are responsible for converting a mixed integer programming problem to a bounded mixed-integer binary (i. e., 0 or 1) programming problem. The PD problem is already in this form therefore we can start directly from Step 3.

The Production problem (lower level) of the B-MILP is transformed to a multi-parametric mixed-integer linear programming (mp-MILP) problem (4.3), in which the optimization variables of the Distribution (upper level) problem, s_{dc} and y_{dc}, are considered as parameters for the lower Production level:

$$
\min_{x_{pd}, z_{pd}} \begin{bmatrix} 20 \\ 25 \\ 20 \\ 25 \end{bmatrix} x_{pd} + \begin{bmatrix} 100 \\ 80 \\ 110 \\ 70 \end{bmatrix} z_{pd}
$$

$$
\begin{aligned}
\text{s.t.} \quad & x_{11} + x_{12} \le 135 \\
& x_{21} + x_{22} \le 100 \\
& -x_{11} + -x_{21} \le -s_{11} - s_{12} - s_{13} \\
& -x_{12} + -x_{22} \le -s_{21} - s_{22} - s_{23} \\
& x_{11} - 100z_{11} \le 0 \\
& x_{12} - 100z_{12} \le 0 \\
& x_{21} - 100z_{21} \le 0 \\
& x_{22} - 100z_{22} \le 0 \\
& -x_{pd} \le 0 \\
& 0 \le s_{dc} \le 35
\end{aligned}
\tag{4.3}
$$

Moving to Step 4, the solution of (4.3) using POP®toolbox [198], results to the complete profile of optimal solutions of the lower Production level problem as explicit functions of the variables of the higher Distribution level problem, with corresponding boundary conditions for different regions in the parametric space (critical regions, CR). The solution is given in Table 4.3 and illustrated through a 2-D plot (s_{12} vs. s_{11}) of the parametric space in Figure 4.2, by fixing four of the parameters, $s_{13}, s_{21}, s_{22}, s_{23}$, at 33.

In Step 5, the computed solutions (Table 4.3) are then substituted into the upper Distribution level problem to formulate six new single-level deterministic mixed-integer linear programming problems, (4.4). More specifically, the expressions for the optimization variables of the lower Production level, x_{pd} and z_{pd}, are substituted in the upper Distribution level in terms of the Distribution optimization variables, s_{pd} and y_{pd}, and the definition of critical regions is added in the upper level as a new set

Table 4.3: Case I: Multiparametric solution of the lower Production level problem.

CR	Definition	Variables
1	$\sum_{d \in D} 0.5774 s_{dc} \leq 57.735 \; \forall d \in D$ $\sum_{c \in C} \sum_{d \in D} -0.4082 s_{dc} \leq -55.1135$	$x_{11} = \sum_{c \in C} s_{1c}, x_{12} = 0, x_{21} = 0,$ $x_{22} = \sum_{c \in C} s_{2c}, z_{11} = 1, z_{12} = 0, z_{21} = 0, z_{22} = 1$
2	$\sum_{d \in D} 0.5774 s_{1c} \leq 57.735$ $\sum_{d \in D} -0.5774 s_{2c} \leq -57.735$ $\sum_{c \in C} \sum_{d \in D} -0.4082 s_{dc} \leq -55.1135$	$x_{11} = 0, x_{12} = \sum_{c \in C} s_{2c},$ $x_{21} = \sum_{c \in C} s_{1c},$ $x_{22} = 0, z_{11} = 0, z_{12} = 1, z_{21} = 1, z_{22} = 0$
3	$\sum_{d \in D} -0.5774 s_{1c} \leq -57.735$ $\sum_{d \in D} 0.5774 s_{2c} \leq 57.735$ $\sum_{c \in C} \sum_{d \in D} -0.4082 s_{dc} \leq -55.1135$	$x_{11} = \sum_{c \in C} s_{1c}, x_{12} = 0,$ $x_{21} = 0, x_{22} = \sum_{c \in C} s_{2c},$ $z_{11} = 1, z_{12} = 0, z_{21} = 0, z_{22} = 1$
4	$\sum_{d \in D} -0.5774 s_{dc} \leq -57.735 \; \forall d \in D$	$x_{11} = \sum_{c \in C} s_{1c}, x_{12} = \sum_{c \in C} s_{2c} - 100, x_{21} = 0,$ $x_{22} = 100, z_{11} = 1, z_{12} = 0, z_{21} = 0, z_{22} = 1$
5	$\sum_{d \in D} 0.5774 s_{2c} \leq 57.735$ $\sum_{c \in C} \sum_{d \in D} 0.4082 s_{dc} \leq 55.1135$	$x_{11} = \sum_{c \in C} s_{1c}, x_{12} = 0, x_{21} = 0,$ $x_{22} = \sum_{c \in C} s_{2c}, z_{11} = 1, z_{12} = 0, z_{21} = 0, z_{22} = 1$
6	$\sum_{d \in D} -0.5774 s_{2c} \leq -57.735$ $\sum_{c \in C} \sum_{d \in D} 0.4082 s_{dc} \leq 55.1135$	$x_{11} = \sum_{c \in C} s_{1c}, x_{12} = \sum_{c \in C} s_{2c},$ $x_{21} = 0, x_{22} = 0, z_{11} = 1, z_{12} = 0, z_{21} = 0, z_{22} = 1$

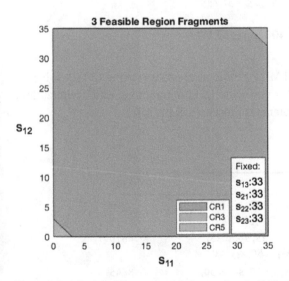

Figure 4.2: 2-D plot of the parametric space given in Table 4.3, at $s_{13} = 33$, $s_{21} = 33$, $s_{22} = 33$ and $s_{23} = 33$.

of constraints:

$$z_1 = \min_{y_{dc}, s_{dc}} \quad 75y_{11} + 60y_{12} + 50y_{13} + 80y_{21} + 30y_{22} + 65y_{23}$$
$$+ 21(s_{11} + s_{12} + s_{13}) + 25(s_{21} + s_{22} + s_{23})$$

$$\text{s.t.} \quad \sum_{d \in D} s_{dc} \geq b_c, \quad \forall c \in C$$

$$s_{dc} \leq 35y_{dc}, \quad \forall c \in C, \quad \forall d \in D$$

$$\sum_{d \in D} 0.5774 s_{dc} \leq 57.735 \quad \forall d \in D$$

$$\sum_{d \in D} -0.5774 s_{2c} \leq -57.735$$

$$\sum_{c \in C} \sum_{d \in D} -0.4082 s_{dc} \leq -55.1135$$

$$\vdots$$

$$z_6 = \min_{y_{dc}, s_{dc}} \quad 75y_{11} + 60y_{12} + 50y_{13} + 80y_{21} + 30y_{22} + 65y_{23}$$
$$+ 21(s_{11} + s_{12} + s_{13}) + 30(s_{21} + s_{22} + s_{23})$$

$$\text{s.t.} \quad \sum_{d \in D} s_{dc} \geq b_c, \quad \forall c \in C$$

$$s_{dc} \leq 35y_{dc}, \quad \forall c \in C, \quad \forall d \in D$$

$$\sum_{d \in D} -0.5774 s_{2c} \leq -57.735$$

$$\sum_{c \in C} \sum_{d \in D} 0.4082 s_{dc} \leq 55.1135$$

(4.4)

For Step 6, the resulting single-level MILP problems are solved using CPLEX solver in GAMS high-level modeling system for mathematical programming and optimization. The solution of the single-level problems is presented Table 4.4.

Table 4.4: Case I: Single-level solutions.

CR	Variables	Objective
1	$s_{11} = 35, s_{12} = 35, s_{13} = 15, s_{21} = 20, s_{22} = 30,$ $s_{23} = 0, x_{11} = 85, x_{12} = 0, x_{21} = 0, x_{22} = 50$	3395
2	$s_{11} = 20, s_{12} = 30, s_{13} = 0, s_{21} = 35, s_{22} = 35,$ $s_{23} = 30, x_{11} = 0, x_{12} = 100, x_{21} = 50, x_{22} = 0$	4610
3	$s_{11} = 35, s_{12} = 35, s_{13} = 15, s_{21} = 20, s_{22} = 30,$ $s_{23} = 0, x_{11} = 85, x_{12} = 0, x_{21} = 0, x_{22} = 50$	3645
4	$s_{11} = 30, s_{12} = 35, s_{13} = 35, s_{21} = 30, s_{22} = 35,$ $s_{23} = 35, x_{11} = 100, x_{12} = 0, x_{21} = 0, x_{22} = 100$	4960
5	$s_{11} = 35, s_{12} = 35, s_{13} = 15, s_{21} = 20, s_{22} = 30,$ $s_{23} = 0, x_{11} = 85, x_{12} = 0, x_{21} = 0, x_{22} = 50$	3330
6	infeasible	infeasible

As a final step the solutions of all the single-level problems are compared. The solution with the minimum objective value (i. e., CR 5) corresponds to the global minimum of the original bilevel programming problem.

4.1.3.2 Case II: Demand uncertainty (mp-B-MILP)

The algorithm used in this paper can be also extended to bilevel problems with uncertainty in one or both optimization levels, making it ideal and, to our knowledge, the only algorithm that can provide the multiparametric solution of PD bilevel problems.

Costumer demand, b_c, only appears in the upper Distribution decision level, therefore Steps 1 to 4 for Case II will be the same with Case I.

Continuing to Step 5, the lower level solutions, corresponding to each critical region, are substituted into the higher level problem. The six resulting single-level problems in this case, are in the form of mp-MILP problems, with the uncertainty b_c treated as a parameter for each problem. Therefore, instead of GAMS, POP®toolbox was used for the solution of the single-level mp-MILP problems. Each critical region formed in Step 4 (Table 4.3, Figure 4.2) is now divided into smaller critical regions as another parametric programming problem is solved within them (Table 4.5). Figure 4.3 illustrates the new critical regions created from follower's CR5.

Table 4.5: Case II: New regions created from the single-level solutions.

Follower's CR	Number of new CRs
CR1	infeasible
CR2	18
CR3	10
CR4	1
CR5	6
CR6	infeasible

The solution of the multiparametric bilevel problem is defined by a linear function of the parameters b_c, capturing the effect of demand on the integrated PD decision problem (solution for follower's CR5 is presented in Table 4.6).

4.1.3.3 Case study conclusion

In this work, we were able to formulate and solve a Production–Distribution planning problem as a bilevel mixed integer linear programming problem. We considered two different cases: (i) constant, known costumer demand, and (ii) uncertain costumer demand. Through the use of the proposed algorithm, we were able to capture the dependence of the integrated Production--Distribution planning decisions on the ever changing, and usually unpredictable, customer's demand.

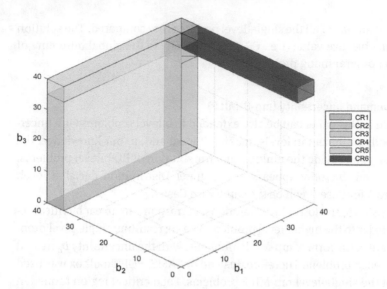

Figure 4.3: 3-D plot of the parametric space created by Follower's CR5.

Table 4.6: Solution for follower's CR5.

CR	Production obj.	Distribution obj.
CR5.1	$25(b_1 + b_2 + b_3) - 335$	$25(b_1 + b_2 + b_3) + 540$
CR5.2	$20(b_1 + b_2 + b_3) + 170$	$21(b_1 + b_2 + b_3) + 485$
CR5.3	$20(b_1 + b_3) + 25b_2 - 5$	$25b_2 + 21(b_1 + b_3) + 475$
CR5.4	$20b_1 + 25(b_2 + b_3) - 180$	$21b_1 + 25(b_2 + b_3) + 500$
CR5.5	$20b_3 + 25(b_1 + b_2) - 180$	$21b_3 + 25(b_1 + b_2) + 515$
CR5.6	$20b_2 + 25(b_1 + b_3) - 180$	$21b_2 + 25(b_1 + b_3) + 550$

4.2 mp-B-QP: Hierarchical model predictive control

As introduced in Chapter 1, hierarchical control structures consist of a hierarchy of control levels, where every level is controlling a subset of the overall control variables, by manipulating a subset of the overall manipulated variables [37–39].

Most attempts to solve hierarchical control problems involve some type of decomposition of the system, and coordination of the decomposed control layers, usually through continuous exchange of data between the optimizers within the same control levels and between different control levels. This can therefore result in suboptimal controllers [38]. Sparse attempts to solve hierarchical control problems using bilevel programming include a consideration of an unconstrained model-based generalized predictive controller in Katebi and Johnson [250], and more recently in Faisca et al. [239], when the idea of using a multiparametric bilevel programming algorithm to solve hierarchical or decentralized MPC structures for the development of explicit MPC controllers was proposed.

In this section, we use the algorithms presented in Chapter 2 for the solution of hierarchical control problems. The proposed methodology will be illustrated through a two-level hierarchical control of a continuous stirred tank reactor (CSTR) system, with an economic objective function at the higher control level, and a set-point tracking controller at the lower control level [251].

4.2.1 CSTR process system

Consider a nonisothermal CSTR, where a first-order exothermic reaction takes place and converts the reactant A to the desired product B (Figure 4.4(a)). The reactant is fed to the reactor through a feedstock stream at concentration C_{A0}, flowrate F, and temperature T_0. The CSTR is assumed to have a constant liquid hold-up. A jacket provides energy to or from the reactor at a heat exchange rate of U.

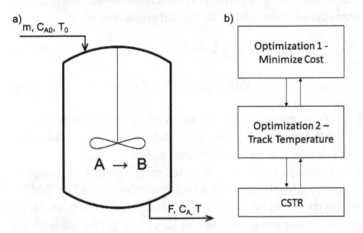

Figure 4.4: (a) Schematic representation of the CSTR process system considered, (b) Two-level control structure of the CSTR system.

For product quality and safety reasons, the temperature of the CSTR is to be kept at around 300 K, with a minimum bound of 285 K, i. e., 5 % set-point violation. The inlet concentration (C_{A0}) and temperature (T_0) are varying between 9 to 10 mol/L, and 200 to 350 K, respectively, and are considered as measured disturbances. The inlet mass flowrate (m) and the coolant temperature (T_c) can be manipulated by the control system, and are therefore considered as manipulated variables.

4.2.2 Controller implementation – PAROC® framework and B-POP

In order to implement the control system for the CSTR described previously, PAROC framework [237] was followed and adapted for hierarchical MPC (Figure 4.5). More de-

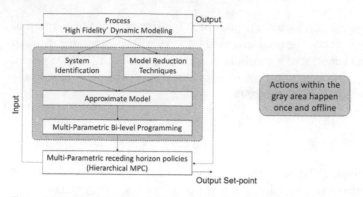

Figure 4.5: PAROC® framework [237] adapted for Hierarchical MPC.

tails on PAROC framework can be found in Appendix C. As a first step, a high fidelity model (4.5) and (4.6) was developed by applying first principles and standard modeling assumptions (constant density and heat capacity, Arrhenius rate, etc.):

$$\frac{dC_A}{dt} = \frac{m}{\rho V}(C_{A0} - C_A) - k_0 C_A e^{-E/RT} \tag{4.5}$$

$$V\rho C_p \frac{dT}{dt} = mC_p(T_0 - T) + V\Delta H_r C_A k_0 e^{-E/RT} + U(T_c - T) \tag{4.6}$$

The above model (4.5)–(4.6) was reduced in complexity via the use of the System Identification Toolbox of MATLAB, to a discrete-time, linear state-space model, which can be further used for the development of receding horizon policies.

Three different control strategies were implemented on the system for comparison purposes. The first one is a bilevel Economic Model Predictive Controller (E-MPC) with set point tracking. This controller has two levels of objectives. The first level is an economic objective and the second level is a classical MPC objective for the tracking of the outlet temperature set-point.

Upper level objective (economic)

$$\min_{T_c} \quad \text{OperatingCost} + \text{SeparationCost} - \text{ProfitC}_B \tag{4.7}$$

Lower level objective (tracking)

$$\min_{m} \quad \sum_{k=1}^{OH-1} \left(QR_k (T_k - T_k^R)^2 \right) \tag{4.8}$$

The second controller is an explicit MPC (mp-MPC) controller with the same objective as the second level of the bilevel controller (4.8), i. e., tracking of the temperature set-point.

Finally, the third controller is an economic explicit MPC (mp-eMPC) controller with the same objective as the first level of the bilevel controller (4.7), i. e., economic.

All controllers have the same set of constraints and are designed following standard multiparametric MPC techniques [252], in MATLAB®, using the algorithms presented in this section and POP® toolbox. More information on the design and implementation of multiparametric MPC as part of PAROC framework is presented in Appendix C. The bilevel controller was solved using the algorithm of Faísca et al. [99], whereas the classical and economic controllers were solved using the mp-QP algorithm implemented in POP® toolbox. The parameters for all controller problems consist of the initial values of the states, the measured disturbances, the previous control action and the output set-point. The control horizon and the prediction horizon for all controllers were set as 1 and 2, respectively, and the discretization time step was set at 1 second.

The size of the resulting programming problems is summarized in Table 4.7. It can be observed that the bilevel controller results into more critical regions than the single-level mp-MPC controllers.

Table 4.7: Size of the resulting multiparametric control problems.

	Bilevel (B-eMPC)		Classical	Economic
	Level 2	Final	(mp-MPC)	(mp-MPC)
Variables	1	1	2	2
Parameters	8	7	7	7
Critical regions	6	29	6	7

The last step of the PAROC Framework is the closed-loop in-silico validation, i. e., the testing of the controllers against the original high fidelity model to verify the performance and robustness of the controllers and tune them appropriately if needed. The inputs and results of the closed loop validation are presented in Figures 4.6, 4.7 and 4.8, and Table 4.8.

The closed loop validation results show that both the B-eMPC controller and the mp-MPC controller manage to follow the temperature set-point with small violations. More specifically, the mp-MPC manages to more accurately follow the temperature set-point, with less sum and average set-point violations than the B-eMPC controller, as this is its only objective. Both are able to effectively reject the disturbances in order to maintain the temperature near the predefined set-point. It is also observed that the B-eMPC controller manages to keep the temperature around 305 K with a more economical way giving approximately 8 times greater profit than the classical mp-MPC (Table 4.8). The mp-eMPC controller has a slightly better accumulative profit (2.46 %) than the B-eMPC but cannot keep the temperature at the set-point, with 1.7 times higher average set-point violation than the B-eMPC.

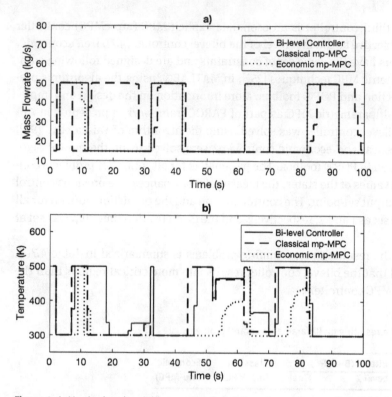

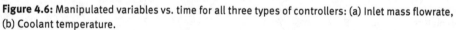

Figure 4.6: Manipulated variables vs. time for all three types of controllers: (a) Inlet mass flowrate, (b) Coolant temperature.

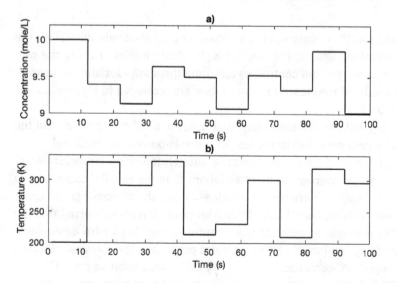

Figure 4.7: Process Disturbances vs. Time: (a) Inlet concentration, (b) Inlet temperature.

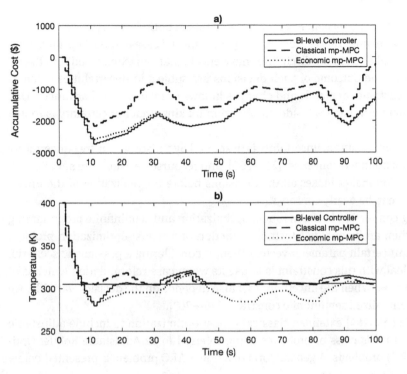

Figure 4.8: Process output vs. Time for all three types of controllers: (a) Accumulative Cost, (b) Reactor temperature.

Table 4.8: Closed loop validation results.

	Classical (mp-MPC)	Economic (mp-eMPC)	Bi-level (B-eMPC)
Accumulative profit ($)	151.78	1194.54	1165.89
$\sum_{k=1}^{100} |T_k - T_k^R|$ (K)	499.40	1283.28	740.19
Average $|T_k - T_k^R|$ (K)	4.94	12.71	7.33

The resulting hierarchical controller is able to effectively reject disturbances and maintain the system at the given set-points in a more economical way than a mp-MPC controller. It was shown that the B-eMPC controller performed more economically than the mp-MPC without significant set-point violations as the mp-eMPC.

4.3 T-MILP: Adjustable robust optimization

Mixed-integer trilevel programming can be applied to many and diverse problems that require hierarchical decision making such as safety and defense [40–42], supply chain management [43], energy management [44] and robust optimization [45–47]. Here, we will focus on applying trilevel programming for a special case of robust optimization.

Decisions taken in many disciplines have effects that can extend well into the future, therefore, the outcome of such decisions are subject to uncertainties, such as variation in customer demands and changes in laws or technological advances. One of the dominant approaches to address decision making under uncertainty is robust optimization.

In robust optimization, uncertainty is described by a distribution-free uncertainty set, typically a bounded convex set [253–260], and recourse decisions are not allowed, i. e., the decision maker makes all the decisions before the realization of the uncertainty, which can be overly conservative.

A strong connection between robust optimization and semiinfinite programming exists, as when an inequality constraint function of a robust optimization problem depends on uncertain parameter vectors then, if considering a pessimistic scenario, the way to deal with this constraint is to use its worst case reformulation which is of semiinfinite type. When the uncertainty set also depends on the decision variable, we arrive at a generalized semiinfinite constraint [261–263].

Ben-Tal et al. [264] extended classical robust optimization to include adjustable decisions, with this class of problems being referred to as Adjustable Robust Optimization (ARO) problems. A general form of a linear ARO problem is presented below (4.9):

$$
\begin{aligned}
\min_{x} \quad & c^T x + \max_{u \in U} \ \min_{y \in \Omega(x,u)} b^T y \\
\text{s. t.} \quad & Ax \geq d, x \in S_x \\
& \Omega(x,u) = \{ y \in S_y : Wy \geq h - Tx - Mu \}
\end{aligned}
\tag{4.9}
$$

where u is the uncertainty set, x are "here-and-now" decisions, i. e., are to be made before the realization of the uncertainty, and y are "wait-and-see" decisions, i. e., are to be made after the realization of the uncertainty.

The uncertainty considered in ARO problems of the form (4.9) is unstructured but bounded as in robust optimization problems. One can easily observe the connection of ARO with two-stage stochastic optimization, were the uncertainty can be modeled as a random vector with a known probability distribution.

ARO problems are very challenging to be solved, with even the simplest linear continuous case being computationally intractable, as discussed in Zhen et al. [265].

A key approach for the approximate solution of ARO problems is to restrict the "wait-and-see" adjustable decisions to be affine functions of the uncertainty [264]. This approach is widely known as affine decision rule approximation and results into computationally tractable problems as the approximated problem can be solved as a static robust optimization problem.

Although ARO has received a lot of attention in the open literature, most of the contributions consider linear continuous problems. A limited number of researchers have tried to tackle the problem of mixed-integer linear ARO problems [266–268], and to our knowledge there are no contributions for the mixed-integer nonlinear ARO problem.

Our key idea for the exact solution of ARO problems came from the observation that the second stage minimization ($\min_{y \in \Omega(x,u)} b^T y$) is multiparametric in terms of the first stage variables (x) and uncertainty (u). Therefore, the idea is to solve the lower level problem multiparametrically considering u and x as parameters, similar to the way we solve trilevel problems.

This step would allow as to arrive to a set of (exact) affine decision rules valid for the whole feasible space of u and "here-and-now" decisions x.

4.3.1 ARO as a trilevel optimization problem

In reality, the "here-and-now" decisions chosen might not always be operational for all realizations of the uncertainty, for example, a processing plant might be more profitable if it is designed for a maximum capacity ("here-and-now" variable) that is less than the demand of some costumers (uncertainty). The formulation of the classical ARO problem does not take this into consideration, therefore, sometimes resulting to an overly conservative solution.

If the uncertainty is defined as a function of "here-and-now" variables (x), instead of being free this would result into problem (4.10), a trilevel optimization problem that can take into consideration "here-and-now" decisions that constrain the feasible space of the uncertainty:

$$
\begin{aligned}
\min_{x} \quad & c^T x + \max_{u \in U(x)} \ \min_{y \in \Omega(x,u)} b^T y \\
\text{s.t.} \quad & Ax \geq d, x \in S_x \\
& \Omega(x,u) = \{y \in S_y : Wy \geq h - Tx - Mu\}
\end{aligned}
\tag{4.10}
$$

Problem (4.10) can be directly solved using B-POP toolbox. This means that we can solve different classes of mixed-integer constrained ARO (MI-C-ARO) problems.

To illustrate the use and benefits of using B-POP to solve the MI-C-ARO problems, three numerical examples are solved using three different computational methods: (a) affine decision rule approximation, (b) column-and-constraint generation algorithm and (c) B-POP.

Column-and-constraint generation algorithm
The basic idea behind this algorithm is to reduce the ARO problem to a single-level optimization problem by enumerating significant extreme points of the polyhedral set U on-the-fly in a decomposition framework [47, 269].

4.3.2 Numerical examples

Example 1: Linear ARO problem

Consider the following problem taken from Ning and You [47]:

$$\min_{x_1, x_2} \quad 3x_1 + 5x_2 + \max_{u \in U_{box}} \min_{y_1, y_2} 6y_1 + 10y_2$$

$$\text{s.t.} \quad x_1 + x_2 \leq 100$$
$$x_1 + y_1 \geq u_1$$
$$x_2 + y_2 \geq u_2 \tag{4.11}$$
$$x_i, y_i \geq 0, \quad i = 1, 2$$
$$U_{box} = \{u_1, u_2 \mid 5.5 \leq u_1 \leq 52.1, 9.5 \leq u_2 \leq 54.8\}$$

The first step is to recast the lowest optimization level into a multiparametric problem, by considering x and u variables as parameters:

$$\min_{y_1, y_2} \quad 6y_1 + 10y_2$$

$$\text{s.t.} \quad x_1 + x_2 \leq 100$$
$$x_1 + y_1 \geq u_1$$
$$x_2 + y_2 \geq u_2 \tag{4.12}$$
$$x_i, y_i \geq 0, \quad i = 1, 2$$
$$U_{box} = \{u_1, u_2 \mid 5.5 \leq u_1 \leq 52.1, 9.5 \leq u_2 \leq 54.8\}$$

Problem (4.12) is then solved using POP® toolbox, to get the optimal parametric solution of y as a set of affine functions of the rest of the optimization variables (x_1, x_2, u_1, u_2) in different critical regions (Table 4.9).

Table 4.9: Parametric solution of problem (4.12).

CR	Definition	Variables
1	$x_1 - u_1 \leq 0$ $x_2 - u_2 \leq 0$ $x_1 + x_2 \leq 100$	$y_1 = -x_1 + u_1$ $y_2 = -x_2 + u_2$
2	$-x_1 + u_1 \leq 0$ $x_2 - u_2 \leq 0$ $x_1 + x_2 \leq 100$	$y_1 = 0$ $y_2 = -x_2 + u_2$
3	$x_1 - u_1 \leq 0$ $-x_2 + u_2 \leq 0$ $x_1 + x_2 \leq 100$	$y_1 = -x_1 + u_1$ $y_2 = 0$
4	$-x_1 + u_1 \leq 0$ $-x_2 + u_2 \leq 0$ $x_1 + x_2 \leq 100$	$y_1 = 0$ $y_2 = 0$

As a next step, four multiparametric problems are formulated, each corresponding to one critical region, by adding the critical region definitions to the set of constraints of the next level problem, substituting in the affine functions for y_1, y_2, and considering the higher level variables (x_1, x_2) as parameters. The four problems created are presented below (4.13)–(4.16).

$$
\begin{aligned}
\min_{u} \quad & -6u_1 - 10u_2 + 6x_1 + 10x_2 \\
\text{s. t.} \quad & x_1 - u_1 \leq 0 \\
& x_2 - u_2 \leq 0 \\
& x_2 + y_2 \geq u_2 \\
& x_1 + x_2 \leq 100 \\
& 5.5 \leq u_1 \leq 52.1 \\
& 9.5 \leq u_2 \leq 54.8
\end{aligned}
\tag{4.13}
$$

$$
\begin{aligned}
\min_{u} \quad & -10u_2 + 10x_2 \\
\text{s. t.} \quad & -x_1 + u_1 \leq 0 \\
& x_2 - u_2 \leq 0 \\
& x_2 + y_2 \geq u_2 \\
& x_1 + x_2 \leq 100 \\
& 5.5 \leq u_1 \leq 52.1 \\
& 9.5 \leq u_2 \leq 54.8
\end{aligned}
\tag{4.14}
$$

$$
\begin{aligned}
\min_{u} \quad & -6u_1 + 6x_1 \\
\text{s. t.} \quad & x_1 - u_1 \leq 0 \\
& -x_2 + u_2 \leq 0 \\
& x_2 + y_2 \geq u_2 \\
& x_1 + x_2 \leq 100 \\
& 5.5 \leq u_1 \leq 52.1 \\
& 9.5 \leq u_2 \leq 54.8
\end{aligned}
\tag{4.15}
$$

$$
\begin{aligned}
\min_{u} \quad & 0 \\
\text{s. t.} \quad & -x_1 + u_1 \leq 0 \\
& -x_2 + u_2 \leq 0 \\
& x_2 + y_2 \geq u_2 \\
& x_1 + x_2 \leq 100 \\
& 5.5 \leq u_1 \leq 52.1 \\
& 9.5 \leq u_2 \leq 54.8
\end{aligned}
\tag{4.16}
$$

The resulting four multiparametric problems are then solved using POP® toolbox. The solution of all of the problems is presented in Table 4.10.

The next step of the algorithm is to substitute the optimal solutions in Table 4.10 into the higher level problem and solve the resulting single-level problems to get four different solution strategies.

Table 4.10: Parametric solution of the middle level problem.

CR	Definition	Variables	Objective
1.1	$0 \leq x_1 \leq 52.1$ $0 \leq x_2 \leq 54.8$ $x_1 + x_2 \leq 100$	$u_1 = 52.1$ $u_2 = 54.8$	$6x_1 + 10x_2 - 860.6$
2.1	$5.5 \leq x_1$ $0 \leq x_2$ $x_1 + x_2 \leq 100$	$u_1 = 5.5$ $u_2 = 54.8$	$10x_2 - 548$
3.1	$0 \leq x_1$ $9.5 \leq x_2$ $x_1 + x_2 \leq 100$	$u_1 = 52.1$ $u_2 = 9.5$	$6x_1 - 312.6$
4	$x_1 + x_2 \leq 100$		0

The final step is to compare those strategies and choose the optimal one to be the exact and global solution of the original trilevel programming problem.

The global solution of the original trilevel problem is $x_1 = 45.2$, $x_2 = 54.8$ and the objective is 451. Note that this solution is the same as the one presented in Ning and You [47].

The optimal affine decision rules, developed through the trilevel multi-parametric programming method are

$$y_1 = -x_1 + u_1$$
$$y_2 = -x_2 + u_2 \tag{4.17}$$

For this ARO problem, using the affine decision rules method results in the same solution as the multiparametric method. The affine decision rules developed through the affine decision rule method are different form the multiparametric method, and are presented below (4.18):

$$y_1 = -0.81438 + 0.14807x_1$$
$$y_2 = 0 \tag{4.18}$$

Example 2: Mixed integer linear ARO problem

Consider ARO problem (4.19) that contains both continuous and integer "here-and-now" optimization variables.

Following three different methods, (i) the affine decision rule method, (ii) the column-and-constraint generation method and (iii) the trilevel method presented here (B-POP) the solutions in Table 4.11 were found.

In this instance, both the column-and-constraint generation algorithm and B-POP return the exact optimal solution whereas the solution generated by the affine decision rules is suboptimal.

Table 4.11: ARO Example 2 Solutions.

	Affine decision rules	Column-and-constraint	B-POP
Objective	0	6,600	6,600
First-stage decisions	$v_1 = 0, v_2 = 0,$ $x_1 = 0, x_2 = 0$	$v_1 = 1, v_2 = 0,$ $x_1 = 24{,}000, x_2 = 0$	$v_1 = 1, v_2 = 0, x_1 = 24{,}000,$ $x_2 = 0$ $y_{11} = -18{,}000\delta_1 + 20{,}000$
Second-stage decision rules	$y_{11} = 0, y_{12} = 0,$ $y_{13} = 0, y_{21} = 0,$ $y_{22} = 0, y_{23} = 0$		$y_{12} = -x_2 - 18{,}000\delta_2 + 20{,}000$ $y_{13} = x_1 + x_2 + 18{,}000\delta_1$ $+\, 18{,}000\delta_2 - 40{,}000$ $y_{21} = 0,\ y_{22} = x_2,\ y_{23} = 0$

B-POP has the capability to also generate the second-stage decision rules for the entire feasible space of the uncertainty and first-stage variables, giving the decision maker more inside into the dynamics of the problem.

$$\max_{x,v} \quad -0.6x_1 - 0.6x_2 - 100{,}000v_1 - 100{,}00v_2 + \min_{\zeta \in U(x,v)} \ \max_{y \in \Omega(x,v,\zeta)} 5.9y_{11}$$
$$+\, 5.6y_{12} + 4.9y_{13} + 5.6y_{21} + 5.9y_{22} + 4.9y_{23}$$

$$\text{s.t.} \quad y_{11} + y_{21} \le \zeta_1$$
$$y_{12} + y_{22} \le \zeta_2$$
$$y_{13} + y_{23} \le \zeta_3$$
$$y_{11} + y_{12} + y_{13} \le x_1$$
$$y_{21} + y_{22} + y_{23} \le x_2$$
$$x_1 \le 130{,}000v_1$$
$$x_2 \le 130{,}000v_2 \tag{4.19}$$
$$x_1, x_2 \ge 0$$
$$v_1, v_2 \in \{0,1\}$$
$$y_{11}, y_{12}, y_{13}, y_{21}, y_{22}, y_{23} \ge 0$$

$$\zeta = \begin{cases} \zeta_1 = 20{,}000 - 18{,}000\delta_1 \\ \zeta_2 = 20{,}000 - 18{,}000\delta_2 \\ \zeta_3 = 20{,}000 - 18{,}000\delta_3 \\ 0 \le \delta_1 \le 1, \quad 0 \le \delta_2 \le 1, \quad 0 \le \delta_3 \le 1 \\ \delta_1 + \delta_2 + \delta_3 \le 2 \end{cases}$$

Example 3: Mixed integer linear ARO problem

Consider ARO problem (4.20) that contains both continuous and integer "here-and-now" optimization variables.

Table 4.12: ARO Example 3 Solutions.

	Affine decision rules	Column-and-constraint	B-POP
Objective	33,680	33,680	30,536
First stage decisions	$x_1 = 1, x_2 = 0,$ $x_3 = 1, z_1 = 458,$ $z_2 = 0, z_3 = 314$	$x_1 = 1, x_2 = 0, x_3 = 1,$ $z_1 = 458, z_2 = 0,$ $z_3 = 314$	$x_1 = 1, x_2 = 0,$ $x_3 = 1, z_1 = 220,$ $z_2 = 0, z_3 = 480$
Second stage decision rules	$y_{11} = 166 + 40g_1 + 40g_2,$ $y_{12} = 0,$ $y_{13} = 220 + 40g_3,$ $y_{21} = 0, y_{22} = 0, y_{23} = 0$ $y_{31} = 40 - 40g_2,$ $y_{32} = 274 + 40g_2, y_{33} = 0$		$y_{11} = 0, y_{12} = 0, y_{13} = z_1$ $y_{21} = 0, y_{22} = z_2, y_{23} = 0$ $y_{31} = 40g_1 + 206$ $y_{32} = -z_2 + 40g2 + 274$ $y_{33} = -z_1 + 40g_3 + 220$

Following the same three different methods, the solutions in Table 4.12 were found.

In this instance, first-stage decision variables can constrain the feasible region of the uncertainty, therefore, enumeration of the extreme points of uncertainty may not result in the optimal solution. For this reason, the column-and-constraint generation algorithm along with the affine decision rules gave a suboptimal solution to the original ARO problem.

$$
\begin{aligned}
\min_{x,z} \quad & 400x_1 + 414x_2 + 326x_3 + 18z_1 + 25z_2 + 20z_3 + \max_{d \in D(x,z)} \min_{y} \quad 22y_{11} \\
& + 33y_{12} + 24y_{13} + 33y_{21} + 23y_{22} + 30y_{23} + 20y_{31} + 25y_{32} + 27y_{33} \\
\text{s.t.} \quad & z_1 \leq 800x_1 \\
& z_2 \leq 800x_2 \\
& z_3 \leq 800x_3 \\
& y_{11} + y_{12} + y_{13} \leq z_1 \\
& y_{21} + y_{22} + y_{23} \leq z_2 \\
& y_{31} + y_{32} + y_{33} \leq z_3 \\
& y_{11} + y_{21} + y_{31} \leq d_1 \\
& y_{12} + y_{22} + y_{32} \leq d_2 \\
& y_{13} + y_{23} + y_{33} \leq d_3 \\
& x_1, x_2, x_3 \in \{0, 1\} \\
& z_1, z_2, z_3 \geq 0 \\
& y_{11}, y_{12}, y_{13}, y_{21}, y_{22}, y_{23}, y_{31}, y_{32}, y_{33} \geq 0 \\
& d = \begin{cases} d_1 = 206 + 40g_1 \\ d_2 = 274 + 40g_2 \\ d_3 = 220 + 40g_3 \\ 0 \leq g_1 \leq 1, \quad 0 \leq g_2 \leq 1, \quad 0 \leq g_3 \leq 1 \\ g_1 + g_2 + g_3 \leq 1.8 \end{cases}
\end{aligned}
$$

(4.20)

4.3.2.1 ARO as a trilevel problem – key overall remarks

Using multiparametric programming to solve the inner problem for "wait-and-see" variables, we were able to show that affine rules are mere approximations, and can be suboptimal for general classes of problems, as different affine decision rules are optimal in different spaces of "here-and-now" and uncertainty variables feasible space.

Moreover, we developed theory for generating exact generalized affine rules for linear/quadratic problems involving both continuous and integer variables in all optimization levels – via the multiparametric programming solution of the lower level optimization problem. Through this, we are able to capture the full set of affine decisions as a function of "here-and-now" decisions and uncertainty.

An important observation is that even if the first-stage objective is changed, the affine decision rules we are able to generate are still valid. Also, the decision maker has access to a plethora of strategies to choose from, opposed to other solution methods that will only generate one strategy.

4.4 Planning and scheduling integration (MF-B-MILP)

In this section, we are considering the simple case study of the planning and scheduling integration of a production processing plant.

4.4.1 Problem description

The problem under investigation considers a production processing plant that produces three different products, A, B and C, through two processing stages. The processing time of product B, for both of the production stages, is proportional to the production target of B, P_B. The processing times of products A and C are constant and are presented in Table 4.13.

Table 4.13: Processing time, T_{jk}.

Product	Stage 1 (hr)	Stage 2 (hr)
A	3	5
B	kP_B	$2k(P_B + 0.5)$
C	6	3

We are considering that one planning period consists of two scheduling periods (Figure 4.9), and at the end of each scheduling period the production facility must supply its customers the demand asked (Table 4.14).

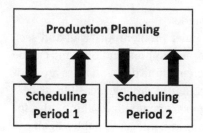

Figure 4.9: Representation of the Production planning and scheduling integration case study problem.

Table 4.14: Costumer demand for each product.

Product	Period 1 (kg)	Period 2 (kg)
A	4	8
B	5	7
C	6	5

4.4.2 Production planning problem

The aim of the production planning decision level is to determine the production target of each product for every scheduling period, so as to minimize the inventory level after each scheduling period comes to an end and make sure that the demand of each product is fulfilled. The formulation of the production problem is presented below (4.21). The notation used is defined in Table 4.15:

$$\min_{P,I} \quad k_A(I_A^1 + I_A^2) + k_B(I_B^1 + I_B^2) + k_C(I_C^1 + I_C^2) + k_4^1 c_{2C}^1 + k_4^2 c_{2C}^2$$

$$\text{s. t.} \quad P_A^1 = D_A^1 - I_A^0 + I_A^1$$

$$P_B^1 = D_B^1 - I_B^0 + I_B^1$$

$$P_C^1 = D_C^1 - I_C^0 + I_C^1$$

$$P_A^2 = D_A^2 - I_A^1 + I_A^2 \tag{4.21}$$

$$P_B^2 = D_B^2 - I_B^1 + I_B^2$$

$$P_C^2 = D_C^2 - I_C^1 + I_C^2$$

4.4.3 Scheduling problems

The objective of each scheduling problem is to minimize the makespan by assigning time slots to each product. The formulation of the scheduling problems is a modification of the formulation developed by Ryu et al. [33] and presented below. The notation

Table 4.15: Notation.

k	Products (A, B, C)
i	Time slot (1, 2, 3)
j	Production stage (1, 2)
n	Scheduling period (1, 2)
P_k^n	Production target for product k in scheduling period n
$T_{j,k}$	Processing time of product k in stage j
D_k^n	Demand of product k in scheduling period n
I_k^n	Inventory level of product k after scheduling period n
$c_{i,j}^n$	Completion time of product k in stage j and period n
$y_{i,k}$	Binary 0-1 variable for product k being assigned in time slot i
CR	Critical region

used is defined in Table 4.15.

$$\min_{y,c} \quad c_{2C}$$

$$\text{s.t.} \quad \sum_{i=1}^{3} y_{i,k} = 1 \qquad \forall k$$

$$\sum_{k=1}^{3} y_{i,k} = 1 \qquad \forall i$$

$$c_{i,j} \geq c_{i,j-1} + \sum_{k=1}^{3} w_{i,k,j} \qquad j > 1, \forall i$$

$$c_{i,j} \geq c_{i-1,j} + \sum_{k=1}^{3} w_{i,k,j} \qquad i > 1, \forall j$$

$$w_{i,k,j} \geq \theta_{k,j} - \theta_{k,j}^{U}(1 - y_{i,k}) \qquad \forall i,j,k$$

$$w_{i,k,j} \leq \theta_{k,j} - \theta_{k,j}^{L}(1 - y_{i,k}) \qquad \forall i,j,k$$

$$y_{i,k}\theta_{k,j}^{L} \leq w_{i,k,j} \leq y_{i,k}\theta_{k,j}^{U} \qquad \forall i,j,k$$

$$\theta_{k,j}^{L} \leq \theta_{k,j} \leq \theta_{k,j}^{U} \qquad \forall i,j,k$$

(4.22)

where i is the time slot $(1, 2, 3)$, j is the production stage $(1, 2)$, and k is the product (A, B, C).

The first constraint ensures that each product is assigned one time slot, whereas the second constrained ensures that just one product is assigned in each time slot. The third and fourth constraint define the completion times of products at different stages.

Nonlinearities arise from the term $w_{i,k,j} = y_{i,k}T_{j,k}$, as $T_{j,B}$ is a function of P_B. The last four constraints are used to eliminate this nonlinearities resulting into a mixed-integer linear programming problem (more information on the reformulation can be found in Ryu et al. [33]).

4.4.4 Solution method

To solve the problem defined above, the algorithm presented in this chapter for bilevel multifollower optimization problems is followed.

Step 1: The two follower problems are reformulated as mp-MILP problems, in which the optimization variables of the upper level problem P^1, P^2 and I^1, I^2 are considered as parameters. Note that the decision variables of each of the follower problems do not appear in the other follower problem and, therefore, are not considered as parameters.

Step 2: The mp-LP problems are then solved using a mp-LP algorithm through POP toolbox [242] and yield the optimal parametric solutions given in Tables 4.16 and 4.17 and presented in Figure 4.10. In this example, each parametric solution consists of only two critical regions. Since the scheduling problems are structurally identical, one could just solve one of the problems and then derive the solution for the rest.

Table 4.16: Multiparametric solution of the first follower – Scheduling period 1.

CR	Definition	Objective-Makespan
1.1	$4 \leq P_B^1 \leq 5.5$	$P_B^1 + 12$
1.2	$5.5 \leq P_B^1 \leq 7$	$2P_B^1 + 6.5$

Table 4.17: Multiparametric solution of the second follower – Scheduling period 2.

CR	Definition	Objective-Makespan
2.1	$4 \leq P_B^2 \leq 5.5$	$P_B^2 + 12$
2.2	$5.5 \leq P_B^2 \leq 7$	$2P_B^2 + 6.5$

Step 3: The solutions obtained are then substituted into the upper level problem to formulate four new single-level deterministic linear programming (LP) problems that correspond to reformulations of the original bilevel multifollower problem. The four problems are created by taking different combinations of the followers critical regions (Table 4.18).

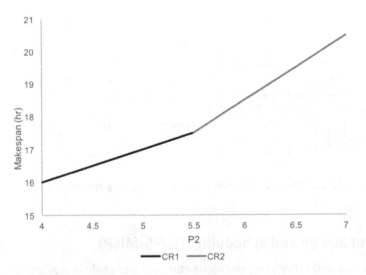

Figure 4.10: Multiparametric solution of the followers problem.

Table 4.18: Combinations of CRs used for the single-level reformulations.

Combination	Critical regions
1	1.1 and 2.1
2	1.1 and 2.2
3	1.2 and 2.1
4	1.2 and 2.2

Table 4.19: Single-level solutions.

Combination	Makespan 1	Makespan 2	Planning obj.
1		infeasible	
2	17.5	19.5	377
3	19.5	17.5	381
4	117.5	19.5	377

Step 4: The single-level problems created in **Step 3** are then solved using CPLEX linear programming solver, and result to the solutions presented in Table 4.19.

After the comparison procedure the global optimum is found to be 377 and lies at the point were critical region 1.1 meets critical region 1.2, resulting to both combinations 2 and 4 giving the same solution. The global optimal solution is also presented in Figure 4.11.

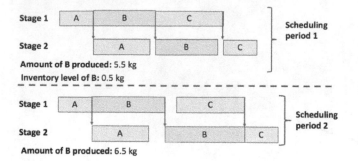

Figure 4.11: Global optimal solution of the original bilevel multifollower planning and scheduling integration problem.

4.5 Integration of design and scheduling (MF-B-MILP)

The integration of design and scheduling decisions can play a big role in designing economically profitable plants and improving their operational performance [270].

Design decisions involve the decisions that must be taken before the plant is operational and are the less likely to change while a possible change usually requires not only a considerable investment but also the permanent cease of operation. Such decisions include the location and capacity of the production plant, the choice of raw materials and products, and the number and capacity of different units in the plant. At the operating level, scheduling decisions optimize the plant performance and involve the detailed timing of operations and sequencing for a fixed process design [271].

In this work, we are focusing on the integration of process design decisions and operation decisions of processing plants. The design and scheduling problem can be expressed as a hierarchical decision problem, where design related decisions occur at the upper level and operational scheduling decisions at the lower level [272, 273]. We formulate this bilevel mixed-integer optimization problem and solve it through a multiparametric bilevel solution algorithm to arrive to the exact global optimum of the integrated problem.

4.5.1 Bilevel formulation for the integration of process design and scheduling

Bilevel formulations have been used extensively in operations research for several years. In this section, we will present a bilevel formulation for the integration of process design and scheduling, where design-related decisions occur at the upper level, and scheduling-related decisions at the lower level optimization problem. The notation used throughout the formulation can be found in Table 4.20.

4.5.1.1 Upper level problem – design

A design problem generally aims at designing a profitable plant by making long term decisions. Those decisions include the location or capacity of the plant, the type of

Table 4.20: Notation.

Sets and indices

i	Time slot $(1,\ldots,N)$
k	Product $(1,\ldots,N)$
j	Stage $(1,\ldots,M)$

Upper level variables

n_j	Integer variable for the number of units in stage j
P_k	Production target of product k
Ca	Plant Capacity

Lower level variables

$c_{i,j}$	Completion time of ith product in stage j
$y_{i,k}$	Binary variable to denote if product k is processed at the ith time slot (sequence)
$w_{i,k}$	Auxiliary variable ($w_{i,k} = y_{i,k}P_k$)

Constant parameters

$A_{j,k}$	Processing time factor of product k in stage j
d_k	Demand of product k
C_j^{InvU}	Unit investment cost for stage j
C_k^{Rev}	Selling price of product k
C_k^{Oper}	Operating cost of product k
P_k^L	Lower bound on the production target of product k
P_k^U	Upper bound on the production target of product k

products it is producing, the pathways to produce these products and the type and number of units needed. Equation (4.23) presents a simplified objective function example of a plant design optimization problem:

$$\min_{n_j,P_k,\text{Ca}} -\sum_{k=1}^{N} C_k^{Rev}P_k + \sum_{k=1}^{N} C_k^{Oper}P_k + \sum_{j=1}^{M} C_j^{InvU}n_j\text{Ca} \qquad (4.23)$$

where the first term corresponds to the revenue gained from selling the products, the second term to corresponds to the operating costs and the final term to the investment costs required to purchase process units. Note that the objective function here is not linear as the last term (investment costs) is bilinear. The design decisions for this example include the choice of the number of units in each processing stage, and the production target of each product.

4.5.1.2 Lower level problem – scheduling

The scheduling problem optimizes the plant performance by determining the detailed timing of operations and sequencing so as to meet a performance criterion, for example minimizing the makespan. The scheduling model generally involves two types of constraints, sequencing constraints that typically denote which products are produced in the different time instances, and assignment constraints that determine the completion times of the products at different stages. The formulation of the scheduling problem (4.24) presented here is a modification of the formulation developed in [33], (also discussed in Section 4.4.3 of this book):

$$\min_{c_{i,j}, y_{i,j}} \quad c_{N,M}$$

$$\text{s. t.} \quad \sum_{i=1}^{N} y_{i,k} = 1 \qquad \forall k$$

$$\sum_{k=1}^{N} y_{i,k} = 1 \qquad \forall i$$

$$c_{i,1} \geq \sum_{k=1}^{N} y_{i,k} A_{j,k} P_k \qquad \forall i \qquad (4.24)$$

$$c_{i,j} \geq c_{i,j-1} + \sum_{k=1}^{N} y_{i,k} A_{j,k} P_k \qquad j > 1, \forall i$$

$$c_{i,j} \geq c_{i-n_j,j} + \sum_{k=1}^{N} y_{i,k} A_{j,k} P_k \qquad i > n_j, \forall j$$

In this formulation, we are assuming that the processing time of each product at each stage is a linear function of the production target, P_k.

The objective function of problem (4.24) is to minimize $c_{N,M}$, that is, the completion time of the last product in the last stage and corresponds to the makespan. The first equality constraint ensures that each product is assigned at one position in the production sequence. The second equality constraint ensures that each position in the sequence is assigned to one product. The third constraint is an inequality constraint and indicates that the completion time of the first stage for all products is greater than the processing time needed. The fourth and fifth constraints indicate that a product in a stage can only be processed if the product and the corresponding unit are available at the same time.

The last three bilinear constraints are linearized by introducing an auxiliary variable, $w_{i,k} = y_{i,k} P_k$, and are updated with the following constraint set (4.25):

$$c_{i,1} \geq \sum_{k=1}^{N} A_{1,k} w_{i,k} \qquad \forall i$$

$$c_{i,j} \geq c_{i,j-1} + \sum_{k=1}^{N} A_{j,k} w_{i,k} \qquad j > 1, \forall i$$

$$c_{i,j} \geq c_{i-n_j,j} + \sum_{k=1}^{N} A_{j,k} w_{i,k} \qquad i > n_j, \forall j \qquad (4.25)$$

$$P_k - P_k^U (1 - y_{i,k}) \leq w_{i,k} \leq P_k - P_k^L (1 - y_{i,k}) \qquad \forall i, j, k$$

$$y_{i,k} P_k^L \leq w_{i,k} \leq y_{i,k} P_k^U \qquad \forall i, k$$

$$P_k^L \leq P_k \leq P_k^U \qquad \forall k$$

The third constraint in the set is only active for $i > n_j$. Since n_j is considered an optimization variable, a reformulation is needed. Big-M constraints are formulated to activate and deactivate this constraint for different values of i and n_j. The integer variable n_j is transformed into a set of binary variables, $m_{j\alpha}$, using the procedure described in Floudas [241] (Section 6.2.1, Remark 1) to allow for the formation of Big-M constraints. One can observe that the design optimization variables, n_j and P_k, appear in the constraints of the scheduling problem. This indicates that solving the two problems separately can result in a suboptimal or even infeasible solution. Therefore, a bilevel formulation and a global solution algorithm for bi-level problems will be able to supply the decision makers with the optimal solution of the design and scheduling problem. The final problem formulation corresponds to a bilevel mixed-integer linear programming problem.

4.5.2 Illustrative case study

Based on the proposed formulation and algorithm, a small case study is solved for illustration purposes. The case study considers the design and scheduling integration of processes that consists of two stages (a reaction stage and a separation stage, Figure 4.12) for the production of three products (A, B and C). At the design phase, the number of units for each stage is decided along with the capacity or production target of the processing plant. At the operating stage, scheduling decisions are made that include the sequence of the production of the three products and the start and finish times of each production stage for each product. The constants used for this case study are presented in Table 4.21, Table 4.22 and Table 4.23.

The maximum number of units for both of the production stages is set to three (Figure 4.12) as the number of products being produced is three. Furthermore, bounds are set for the maximum and minimum production capacity of the three products and this are set to 20 and 10 tons, respectively.

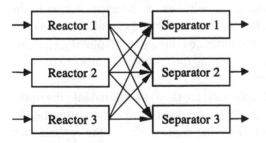

Figure 4.12: A schematic representation of the process configuration of the illustrative example.

Table 4.21: Processing time data.

Product, k	Processing time factor, $A_{j,k}$ (h/Ton)	
	Stage 1 ($j = 1$)	Stage 2 ($j = 2$)
A	0.10	0.30
B	0.07	0.20
C	0.10	0.25

Table 4.22: Operating cost, demand and selling price data.

Product, k	Operating Cost, C_k^{Oper} ($/Ton)	Demand, d_k (Ton)	Price, C_k^{Rev} ($/Ton)
A	30	20	600
B	33	19	720
C	27	17	880

Table 4.23: Unit investment cost data.

Stage, j	Cost, C_j^{InvU} ($/Ton)
Reactor, $j = 1$	300
Separator, $j = 2$	600

4.5.2.1 Solution method

Following the algorithm presented Table 2.1 and Section 2.1, the first and second steps are skipped as the problem is already a binary B-MILP. For the third step, the lower level scheduling problem, (4.25) is solved as a multiparametric problem where the design decisions, number of units (n_j), production target of each product (P_k) and production capacity (Ca), are considered as parameters. The solution of the multiparametric problem resulted into the complete profile of optimal solutions of the lower scheduling level problem as explicit functions of the variables of the higher design level problem, with corresponding boundary conditions for different regions in the parametric space (critical regions, CR). The solution consists of 25 critical regions and a fraction of them is given in Table 4.24 and illustrated through a 3-D plot (P_A vs. P_B vs. P_C) of the parametric space in Figure 4.13, by fixing the number of units (n_j) to $n_1 = 1$ and $n_2 = 3$ (one reactor and 3 separators).

In Step 5, the computed solutions (Table 4.24) are then substituted into the upper design level problem to formulate new single-level deterministic mixed-integer bilinear programming problems. More specifically, the expressions for the optimization variables of the lower scheduling level, $c_{(i,j)}$ and $y_{(i,j)}$, are substituted in the upper design level in terms of the design optimization variables, n_j and P_k, and the definition of critical regions is added in the upper level as a new set of constraints.

Table 4.24: Partial solution of the lower level scheduling problem.

CR	Definition of the CR	Lower level variables
C1	$0.359P_A + 0.252P_B - 0.899P_C \leq 0$ $-0.926P_A + 0.216P_B - 0.309P_C \leq 0$ $-0.408P_A + 0.816P_B - 0.408P_{\leq}0$ $10 \leq P_A \leq 20$ $P_B \geq 10$ $10 \leq P_C \leq 20$ $n_1 = 1, n_2 = 3$	$c_{1,1} = 0.07P_A$ $c_{1,2} = 0.27P_B$ $c_{2,1} = 0.1P_A + 0.07P_B$ $c_{2,2} = 0.4P_A + 0.07P_B$ $c_{3,1} = 0.1P_A + 0.07P_B + 0.1P_C$ $c_{3,2} = 0.1P_A + 0.07P_B + 0.35P_C$ $y_{1,A} = 0, y_{1,B} = 1$ $y_{1,C} = 0, y_{2,A} = 1$ $y_{2,B} = 0, y_{2,C} = 0$ $y_{3,A} = 0, y_{3,B} = 0$ $y_{3,C} = 1$
C2	$0.359P_A + 0.252P_B - 0.899P_C \leq 0$ $-0.926P_A + 0.216P_B - 0.309P_C \leq 0$ $-0.408P_A + 0.816P_B - 0.408P_{\leq}0$ $10 \leq P_A \leq 20$ $P_B \geq 10$ $10 \leq P_C \leq 20$ $n_1 = 1, n_2 = 3$	$c_{1,1} = 0.07P_A$ $c_{1,2} = 0.27P_B$ $c_{2,1} = 0.1P_A + 0.07P_B$ $c_{2,2} = 0.4P_A + 0.07P_B$ $c_{3,1} = 0.1P_A + 0.07P_B + 0.1P_C$ $c_{3,2} = 0.27P_B + 0.25P_C$ $y_{1,A} = 0, y_{1,B} = 1$ $y_{1,C} = 0, y_{2,A} = 1$ $y_{2,B} = 0, y_{2,C} = 0$ $y_{3,A} = 0, y_{3,B} = 0$ $y_{3,C} = 1$

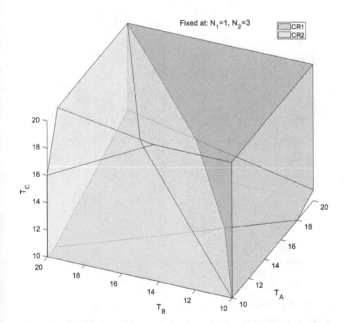

Figure 4.13: 3-D plot of the parametric space for fixed number of units.

For Step 6, the resulting single-level MIP problems are solved using CPLEX algorithm. The solution of a fraction of the single-level problems created is presented Table 4.25.

Table 4.25: Single-level solutions of a fraction of the CRs created in Step 4.

CR	Upper level objective function	Lower level objective function	Production sequence	Fraction of the variables
A1	−21248	15.080	B-A-C	$n_1 = 1,$ $n_2 = 1,$ $P_A = 19,$ $P_B = 19,$ $P_C = 17$
A2	−21284	15.080	B-C-A	$n_1 = 1,$ $n_2 = 1,$ $P_A = 19,$ $P_B = 19,$ $P_C = 17$
B1	−10370	8.840	B-A-C	$n_1 = 1,$ $n_2 = 2,$ $P_A = 17,$ $P_B = 17,$ $P_C = 17$
B2	−10370	9.350	C-A-B	$n_1 = 1,$ $n_2 = 2,$ $P_A = 17,$ $P_B = 17,$ $P_C = 17$
		...		
C1	−170	9.350	C-A-B	$n_1 = 1,$ $n_2 = 3,$ $P_A = 17,$ $P_B = 17,$ $P_C = 17$
C1	−170	8.840	B-A-C	$n_1 = 1,$ $n_2 = 3,$ $P_A = 17,$ $P_B = 17,$ $P_C = 17$
		...		

As a final step, the solutions of all the single-level problems are compared. The solution with the minimum objective function value corresponds to the global minimum of the original bilevel programming problem. For this case, the optimum lies in CR

A1 and CR A2 that result to the same upper and lower objective functions but have different sequence for the production of the three products. The upper level objective is \$ −21284 and the lower level objective is 15.08 hr. The optimal design variables are $P_A = 19$, $P_B = 19$, $P_C = 17$, Ca = 19, $n_1 = 1$, $n_2 = 1$. The optimal sequence of production is either B-A-C or B-C-A. It is worth noting here that optimistic, pessimistic and degenerate solutions can be found using the proposed methodology, supplying the decision maker with all optimal solutions.

In this case study, we were able to formulate and solve a design and scheduling integration problem as a bilevel mixed integer programming problem. Through the algorithm described in Chapter 2 we were able to get the global solution of the bilevel problem that considered both design and operational decisions.

4.5.3 Conclusion

This chapter introduced different application case studies where multilevel optimization formulations can or should be utilized. Along with the formulation of the problems, this chapter also presented the solution procedure for each case study.

The next chapter introduces the B-POP toolbox that was used to solve most of the application case studies covered in this chapter.

5 Computational implementation – B-POP toolbox

Considering that mixed-integer bilevel programming problems are very challenging to solve, publicly available bilevel toolboxes are very limited.

Motivated by the lack of an available toolbox for the solution of bilevel mixed-integer programming problems, this chapter presents B-POP®, a MATLAB® based toolbox for the global solution of different classes of bilevel programming problems using multiparametric programming algorithms [274]. The toolbox is an extension to POP® (Parametric OPtimization) toolbox [242], and features (i) bilevel programming solvers for linear and quadratic programming problems and their mixed-integer counterparts, (ii) a versatile problem generator capable of creating random bilevel problems of arbitrary size and (iii) a library of bilevel programming test problems.

The toolbox is based on multiparametric theory [274, 240] and our developed algorithms presented in Chapter 3 [99, 243] and software tools [242].

5.1 Bilevel programming solvers

In B-POP, we have implemented the algorithm of Faisca et al. [99] for B-LP and B-QP problems and the algorithm of Avraamidou et al. [243] for B-MILP and B-MIQP problems. Both algorithms have been combined in a single wrapper:

Solution = BPOP(problem)

where *problem* is the structured array containing the B-LP/B-QP/B-MILP or B-MIQP problem.

The user has the option to choose which LP and QP solver to use (MATLAB [in-built functions linprog and quadprog], NAG and CPLEX), which mp-LP/mp-QP POP solver to use (Geometrical [180], Combinatorial [186], Graph [191] or MPT [193]) and the integer handling technique for B-MIP. This features can be set through the OptionSet function, which also contains all the adjustable settings of POP (see Oberdieck et al. [242] or POP User Manual available at *parametric.tamu.edu/POP* for more details).

5.2 Bilevel problem generator

The aim of this feature is to generate random, feasible bilevel problems of the form (5.1):

https://doi.org/10.1515/9783110760316-005

$$\min_{x_1, y_1} \quad \left(Q_1^T \omega + c_1\right)^T \omega + c_{c1}$$

$$\text{s. t.} \quad A_1 x + E_1 y \leq b_1$$

$$\min_{x_2, y_2} \quad \left(Q_2^T \omega + c_2\right)^T \omega + c_{c2}$$

$$\text{s. t.} \quad A_2 x + E_2 y \leq b_2 \tag{5.1}$$

$$x = [x_1^T \ x_2^T]^T, \quad x \in \mathbb{R}^n$$

$$y = [y_1^T \ y_2^T]^T, \quad y \in \{0, 1\}^p$$

$$\omega = [x^T \ y^T]^T$$

where $Q_1, Q_2 \in \mathbb{R}^{n \times n} > 0$, $c_1, c_2 \in \mathbb{R}^n$, $A_1 \in \mathbb{R}^{m_1 \times n}$, $A_2 \in \mathbb{R}^{m_2 \times n}$, $b_1 \in \mathbb{R}^{m_1}$, $b_2 \in \mathbb{R}^{m_2}$ and x_1 is compact (closed and bounded).

The algorithm of the generator can be decomposed in three steps:

Step 1 – Level 1 objective function: In order to define the first level objective function, Q_1, c_1 and c_{c1} need to be defined. Q_1 needs to be symmetric positive definite the size of $\omega \times \omega$, and this is achieved by randomly generating a diagonal matrix featuring only positive entries. For the definition of c_1 and c_{c1}, no specific criterion is required, therefore, a random vector the size of x is generated for c_1, and a random scalar is generated for c_{c1}.

Step 2 – Level 2 objective function: The same strategy followed for the first level objective function is also followed for the second level objective function. Q_2 needs to be a symmetric positive definite matrix the size of $\omega \times \omega$, and this is achieved by randomly generating a diagonal matrix featuring only positive entries. For the definition of c_2 and c_{c2}, similar to Level 1, no specific criterion is required.

Step 3 – Constraints: For the generation of constraints for the bilevel optimization problems, we follow the criteria we developed for the generation of constraints for multiparametric programming problems [242]. These are (i) feasibility and (ii) tightness in the sense that different solutions should be optimal in different parts of the parameter space. The algorithm used to define this constraints was presented in Oberdieck et al. [242], Algorithm 1.

It is worth mentioning here that there are publicly available B-MILP instance problems generated by Matteo Fischetti, Ivana Ljubic, Michele Monaci and Markus Sinnl in their website "https://msinnl.github.io/pages/bilevel.html." Those problems are not of the general B-MILP form presented in (2.1) as the leader variables that influence the follower decisions can only be an integer.

The bilevel problem generator is accessible from the Command Window as

problem = BilevelProblemGenerator(Type, Size, options)

where Type is "BLP," "BQP," "BMILP" or "BMIQP" and Size is a structured array featuring the desired dimensions of the optimization variables of each optimization level and constraints.

5.3 Bilevel problem library

The final feature of B-POP is a bilevel problem library, featuring problem sets of the four different problem types and subclasses, "BLP," "BQP," "BMILP" and "BMIQP," each containing randomly generated problems of different sizes. These problem libraries are used in Section 5.4 to analyze the performance and scalability of BPOP.

Each problem in the bilevel problem library is stored in the folder named "Bilevel Library," which contains a subfolder for each different test set class, which in return contains all the randomly generated problems as ".mat" files.

5.4 Assessing the efficiency and performance of B-POP

Four sets of bilevel problems of different classes, sizes and structures were solved to show the capabilities of B-POP toolbox. Tables 5.1 to 5.4 present the computational results, where x_T denotes the total number of continuous variables of the bilevel problem, y_T denotes the total number of binary variables of the bilevel problem, x_1 and x_2 denote the number of continuous decision variables of the first and second optimization level, respectively, y_1 and y_2 denote the number of binary decision variables of the first and second optimization level, respectively, m denotes the number of constraints of the lower level problem, **Level 1** and **Level 2** denote the time B-POP takes to solve each optimization level and **Total** denotes the total computational time for each test problem. The computational results are also illustrated in Figure 5.1 for the continuous problems and Figure 5.2 for mixed-integer problems.

The computations were carried out on a 2-core machine with an Intel Core i7 at 3.1 GHz and 16 GB of RAM, MATLAB R2016a and IBM ILOG CPLEX Optimization Studio 12.6.3. Note that the independent problems in Step 5 of the B-QP algorithm, and Step 6 of the B-MIP algorithm, were solved sequentially and not simultaneously (which obviously would have improved the computational performance).

5.5 Discussion on the computational results

The solution of the test set problems highlight the capabilities of B-POP.

For the continuous bilevel problems (B-LP, B-QP), the computational performance of B-POP was shown to be very efficient, especially for the bilevel linear problems, where problems with more than 500 variables could be solved in less than 300 seconds. Bilevel quadratic problems are significantly less efficient with problems with more than 100 variables needing more than 500 seconds to be solved.

For the mixed-integer bilevel problems (B-MILP, B-MIQP), similar to the continuous problems, the efficiency for the linear problems was much better than for the nonlinear quadratic problems.

Table 5.1: Computational results: B-LP.

Problem	x_1	x_2	m	mp-Level 2 (s)	Single level (s)	Total time (s)
test1	2	2	10	0.2939	0.0428	0.3367
test2	2	7	36	0.7889	0.0534	0.8423
test3	2	7	18	0.1776	0.0051	0.1827
test4	2	7	18	0.2488	0.0249	0.2737
test5	10	10	40	0.5160	0.0289	0.5449
test6	12	12	24	1.4423	0.0382	1.4806
test7	15	15	90	0.7352	0.0283	0.7636
test8	17	17	34	0.3994	0.0051	0.4044
test9	19	19	76	63.2557	0.2205	63.4762
test10	20	20	80	1.1779	0.0442	1.2221
test11	22	22	132	1.5016	0.0190	1.5206
test12	25	25	100	1.2853	0.0230	1.3082
test13	27	27	20	0.9264	0.0538	0.9802
test14	30	30	22	0.8055	0.0113	0.8168
test15	35	35	30	0.9586	0.0040	0.9626
test16	40	40	35	2.1567	0.0401	2.1968
test18	50	50	40	2.3846	0.0026	2.3872
test20	55	55	50	2.4196	0.0032	2.4227
test21	60	60	55	2.3239	0.0022	2.3261
test24	70	70	70	3.5598	0.0022	3.5620
test26	80	80	80	7.6918	0.0065	7.6983
test28	90	90	90	5.3004	0.0035	5.3040
test30	100	100	100	15.7899	0.0029	15.7928
test31	110	110	110	13.5997	0.0043	13.6039
test32	120	120	120	41.1437	0.0049	41.1486
test33	130	130	130	55.8976	0.0081	55.9057
test34	140	140	140	55.3011	0.0045	55.3056
test35	150	150	150	39.1210	0.0056	39.1266
test36	160	160	160	46.0358	0.0088	46.0446
test37	170	170	170	55.7919	0.0075	55.7994
test38	180	180	180	71.1564	0.0137	71.1701
test39	190	190	190	105.4210	0.0098	105.4308
test40	200	200	200	100.2806	0.0147	100.2953
test41	220	220	220	184.5859	0.0158	184.6018
test42	240	240	240	151.4938	0.0180	151.5119
test43	260	260	260	289.6922	0.0245	289.7168
test44	280	280	280	225.3628	0.0265	225.3893
test45	300	300	300	1060.1907	0.0344	1060.2251

Table 5.2: Computational results: B-QP.

Problem	x_1	x_2	m	mp-Level 2 (s)	Single level (s)	Total time (s)
test1	2	2	16	0.6593	0.0948	0.7541
test2	3	3	17	0.6159	0.0783	0.6943
test3	5	5	30	53.7702	5.5880	59.3583
test5	2	7	5	0.1453	0.0152	0.1605
test6	3	7	7	0.7189	0.1486	0.8674
test7	4	7	10	2.1774	0.3588	2.5361
test8	5	7	10	3.0904	0.4446	3.5350
test9	6	7	10	2.1511	0.2712	2.4223
test11	4	10	10	5.1546	0.8471	6.0017
test13	6	10	10	16.2266	2.1169	18.3434
test15	8	10	10	30.1686	3.0333	33.2018
test17	10	10	10	39.4925	2.8606	42.3531
test19	12	10	10	47.5784	2.8586	50.4370
test20	14	10	10	57.7805	2.5489	60.3294
test21	16	10	10	70.3945	2.9120	73.3065
test22	16	12	10	75.5874	3.2628	78.8503
test23	16	14	10	104.0171	3.1920	107.2091
test24	16	16	10	78.1582	2.6769	80.8351
test25	18	18	10	83.4105	2.7502	86.1606
test26	20	20	10	97.3432	2.6917	100.0349
test27	22	22	10	65.3891	2.1492	67.5383
test28	24	24	10	146.8141	3.9729	150.7870
test29	26	26	10	152.3434	3.6763	156.0197
test30	28	28	10	161.6853	3.9808	165.6661
test31	30	30	10	164.9478	3.3726	168.3204
test32	32	32	10	196.2318	3.0421	199.2739
test33	34	34	10	208.1788	3.2885	211.4672
test34	36	36	10	252.2779	3.7759	256.0538
test35	38	38	10	306.4905	3.3287	309.8192
test36	40	40	10	289.6366	4.4620	294.0987
test37	42	42	10	313.0398	4.7300	317.7698
test38	44	44	10	341.7009	5.9530	347.6539
test39	46	46	10	388.4117	4.4595	392.8712
test40	48	48	10	417.8896	4.9598	422.8495
test41	50	50	10	441.9685	4.8927	446.8612
test42	52	52	10	546.9051	5.5651	552.4702
test44	54	54	10	512.1658	4.9903	517.1562
test45	56	56	10	637.0739	5.2642	642.3381

Table 5.3: Computational results: B-MILP.

Problem	x_1	y_1	x_2	y_2	m	mp-Level 2 (s)	Single level (s)	Total time (s)
test5	25	25	2	2	15	321.5283	0.9317	322.4600
test6	30	30	2	2	17	107.7841	0.4679	108.2520
test8	40	40	2	2	22	291.130	0.1731	291.3035
test10	50	50	2	2	27	455.0280	0.9539	455.09819
test14	2	2	20	20	12	0.6848	0.0016	0.6864
test16	2	2	30	30	17	1.5148	0.0031	1.5179
test18	2	2	40	40	22	1.2439	0.0021	1.2460
test19	2	2	45	45	25	1.5815	0.0027	1.5842
test20	2	2	50	50	27	1.7802	0.0023	1.7825
test21	2	2	55	55	30	1.6793	0.0019	1.6812
test22	2	2	60	60	32	1.9333	0.0020	1.9353
test23	2	2	65	65	35	3.1148	0.0020	3.1168
test24	2	2	70	70	37	2.7563	0.0026	2.7589
test25	2	2	75	75	40	4.1390	0.0026	4.1416
test26	2	2	80	80	42	3.8449	0.0023	3.8472
test27	2	2	85	85	45	5.6016	0.0031	5.6047
test28	2	2	90	90	47	5.4821	0.0032	5.4853
test29	2	2	95	95	50	5.9115	0.0029	5.9144
test30	2	2	100	100	52	8.1234	0.0029	8.1263
test33	10	10	10	10	10	26.1063	0.0233	26.1296
test37	20	20	20	20	20	4.3209	0.0102	4.3312
test41	30	30	30	30	30	3.5103	0.0019	3.5122
test44	40	40	40	40	40	5.4329	0.0025	5.4354
test45	45	45	45	45	45	4.3939	0.0021	4.3960
test46	50	50	50	50	50	8.3232	0.0099	8.3331
test47	55	55	55	55	55	8.9885	0.0034	8.9918
test48	60	60	60	60	60	17.5849	0.0109	17.5957
test49	65	65	65	65	65	10.2242	0.0046	10.2288
test50	70	70	70	70	70	24.5837	0.0123	24.5960
test51	75	75	75	75	75	18.4030	0.0057	18.4087
test52	80	80	80	80	80	13.8105	0.0069	13.8175
test53	85	85	85	85	85	19.7149	0.0142	19.7291
test54	90	90	90	90	90	37.1772	0.0080	37.1852
test55	95	95	95	95	95	59.2469	0.0159	59.2628
test56	100	100	100	100	100	55.3647	0.0080	55.3727
test57	105	105	105	105	105	45.3738	0.0095	45.3833
test58	110	110	110	110	110	68.0360	0.0074	68.0434
test59	115	115	115	115	115	100.8653	0.0119	100.8772
test60	120	120	120	120	120	191.6486	0.0477	191.6963

Table 5.4: Computational results: B-MIQP.

Problem	x_1	y_1	x_2	y_2	m	mp-Level 2 (s)	Single level (s)	Total time (s)
test1	5	5	2	2	5	4.1001	0.1238	4.2239
test2	10	10	2	2	7	2.6959	0.0377	2.7336
test3	15	15	2	2	10	152.1813	0.4648	152.6460
test4	20	20	2	2	12	201.1591	0.5662	201.7052
test5	25	25	2	2	15	175.0922	0.9555	176.0477
test6	2	2	5	5	5	79.3080	0.0730	79.3742
test7	2	2	10	10	7	257.4909	0.0060	257.4969
test8	5	2	5	2	3	8.5233	0.1221	8.6353
test9	10	2	10	2	5	33.0615	0.0601	33.6625
test10	15	2	15	2	6	32.3564	0.0532	32.4096
test11	20	2	20	2	7	47.3702	0.5851	47.9553
test12	25	2	25	2	8	4.5226	0.0345	4.5571
test13	20	5	20	2	5	165.4143	0.7007	166.1150
test14	10	10	30	5	1	210.9560	0.0386	210.9946
test15	5	2	25	5	1	183.3612	0.0856	183.4468

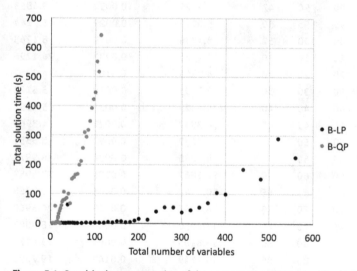

Figure 5.1: Graphical representation of the computational time required to solve continuous bilevel linear and quadratic problems in B-POP.

The number of constraints was also a key factor for the difficulty of each test problem, and it was shown, especially in Figure 5.1, that by increasing the number of constraints the time required to solve the bilevel problems is increased.

Approximate solution methods could be computationally more efficient than the solvers in B-POP, but B-POP arrives to the exact solution of bilevel problems paying a

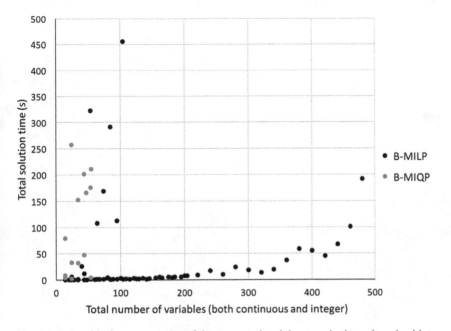

Figure 5.2: Graphical representation of the computational time required to solve mixed-integer bilevel linear and quadratic problems in B-POP.

penalty in computational time for the solution of the mutliparametric programming problems. The number of single-level problems required to be solved does not play a big role in the total computational time as for all four classes of problems solved, and it was clearly observed that the limiting step of the algorithms, and more time consuming, is the solution of the lower level multi-parametric problem.

A Appendix

A.1 PARametric optimization and control framework

The presence of uncertainty in process systems is one of the key reasons for deviation from set operation policies. As these uncertainties realize themselves on different time scales such as on a control, scheduling or design level, an integrated, comprehensive approach to consider uncertainty is required. To address this, the PAROC (PARametric Optimization and Control) framework was developed [275]. PAROC, a novel unified framework for the design, operational optimization and advanced model-based control of process systems, which decomposes this challenging problem into a series of steps are summarized in this Appendix.

PAROC is a comprehensive framework that enables the representation and solution of demanding model-based operational optimization and control problems following an integrated procedure featuring high-fidelity modeling, approximation techniques and optimization-based strategies, including multiparametric programming. A step-by-step description of the framework is provided below and illustrated in Figure A.1. The full description of the framework, as well as its principles are presented in detail in [275].

Step 1: "High fidelity" dynamic modeling
The development of the "high fidelity model," its quality and robustness determine the validity of the framework. The modeling of the system takes place in gPROMS® [276].

Step 2: Model approximation
The resulting highly complex dynamic models of the subsystems of the first step (most commonly DAE or PDAE programs), although sufficiently accurate compared to the real process, are not directly suitable for multiparametric programming studies. Hence, reduction techniques ([277] and [278]) and identification methods (System Identification Toolbox of MATLAB®) are employed to (i) reduce the model complexity while (ii) preserving the model accuracy.

Step 3: Design of the multiparametric model predictive controllers
The design of the controllers is based on the validated procedure described in [279] and [252]. The resulting multiparametric program is solved via the POP® toolbox in MATLAB®, thus acquiring the map of optimal control actions.

Step 4: Closed-loop validation
The procedure is validated through a closed-loop procedure, where the controllers are tested against the original model of step 1. This can happen either via the interoper-

https://doi.org/10.1515/9783110760316-006

ability between software tools such as gPROMS® and MATLAB® via gO:MATLAB or via the straight implementation of the controllers in the gPROMS® simulation via the use of C++ programming and the creation of dynamic link libraries.

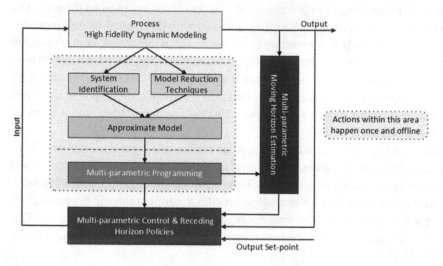

Figure A.1: The PAROC framework approach. Actions within the gray area happen once and offline.

B Appendix

B.1 POP – The parametric optimization toolbox

With an ever increasing number of applications for multiparametric programming, there is a high need for a comprehensive software tool capable of efficiently solving multiparametric programming problems, while being easily embedded into other software architectures such as the ones used in the PAROC platform. Thus, within our group we have developed POP, the parametric optimization toolbox featuring:

- a state-of-the-art multiparametric programming solver for continuous and mixed-integer problems
- a comprehensive problem library featuring an ever increasing number of example problems
- a versatile problem generator, which allows for the generation of random multiparametric programming problems of arbitrary size
- a link to the software YALMIP

In addition, POP is equipped with a graphical user interface, which enables the user-friendly use of all functionalities of POP and a link to the solvers of the Multiparametric Toolbox (MPT), as well as the ability to design explicit MPC problems, through PAROC framework discussed in Appendix A. The interface of POP toolbox is presented in figure B.1.

In general, POP considers the following optimization problem:

$$
\begin{aligned}
z^*(\theta) = \min_{x,y} \quad & (Q\omega + H_t\theta + c)^T\omega + (Q_t\theta + c_t)^T\theta + c_c \\
\text{s.t.} \quad & Ax + Ey \leq b + F\theta \\
& A_{eq}x + E_{eq}y = b_{eq} + F_{eq}\theta \\
& x \in \mathbb{R}^n, \quad y \in \{0,1\}^p, \quad \omega = [x^T \quad y^T], \\
& \theta \in \Theta \subset \mathbb{R}^q \mid CR_A\theta \leq CR_b
\end{aligned}
\tag{B.1}
$$

where $Q \prec 0$, the matrices have appropriate dimensions and which is referred to as a multiparametric mixed-integer quadratic programming (mp-MIQP) problem, as well as its simpler counterparts.

For further reading on the POP toolbox and how to download and use it, the reader is directed to:

- Oberdieck, R.; Diangelakis, N. A.; Papathanasiou, M. M.; Nascu, I.; Pistikopoulos, E. N. POP – Parametric Optimization Toolbox. Industrial and Engineering Chemistry Research 2016, 55 (33), 8979–8991.
- Pistikopoulos, E. N.; Diangelakis, N. A.; Oberdieck, R. Multi-parametric Optimization and Control; John Wiley & Sons, 2020.
- The user manual at https://parametric.tamu.edu/POP/.

https://doi.org/10.1515/9783110760316-007

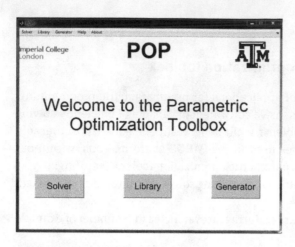

Figure B.1: Interface of the POP toolbox.

C Summary of mp-MILP and mp-MIQP solution algorithms

C.1 Algorithm for the solution of mp-MILP problems

A schematic representation, Figure C.1, showing the steps of the algorithm for the solution of mp-MILP problems (C.1) by [198] is shown below:

$$
\begin{aligned}
\min_{x,y} \quad & Q\omega + H\theta \\
\text{s.t.} \quad & Ax + Ey \leq b + F\theta \\
& x \in \mathbb{R}^n, \quad y \in \{0,1\}^p, \quad \omega = \begin{bmatrix} x^T & y^T \end{bmatrix} \\
& \theta \in \Theta := \{\theta \in \mathbb{R}^q \mid CR_A\,\theta \leq CR_b\},
\end{aligned} \tag{C.1}
$$

where $Q \in \mathbb{R}^{(n+p)\times(n+p)} > 0$, $H \in \mathbb{R}^{(n+p)\times q}$, $A \in \mathbb{R}^{m\times n}$, $E \in \mathbb{R}^{m\times p}$, $b \in \mathbb{R}^m$, $F \in \mathbb{R}^{m\times q}$ and Θ is compact.

C.2 Algorithm for the solution of mp-MIQP problems

The algorithm of [199] for the solution of problems with the general formulation of (C.2) is summarized below. It is based on the decomposition algorithm shown graphically in Figure C.2:

$$
\begin{aligned}
\min_{x,y} \quad & (Q\omega + H\theta + c)^T \omega \\
\text{s.t.} \quad & Ax + Ey \leq b + F\theta \\
& x \in \mathbb{R}^n, \quad y \in \{0,1\}^p, \quad \omega = \begin{bmatrix} x^T & y^T \end{bmatrix} \\
& \theta \in \Theta := \{\theta \in \mathbb{R}^q \mid CR_A\,\theta \leq CR_b\},
\end{aligned} \tag{C.2}
$$

where $Q \in \mathbb{R}^{(n+p)\times(n+p)} > 0$, $H \in \mathbb{R}^{(n+p)\times q}$, $c \in \mathbb{R}^{n+p}$, $A \in \mathbb{R}^{m\times n}$, $E \in \mathbb{R}^{m\times p}$, $b \in \mathbb{R}^m$, $F \in \mathbb{R}^{m\times q}$ and Θ is compact.

Initialization

A candidate solution for the binary variables is found by solving the MIQP problem formed when considering parameters as optimization variables. A binary solution is obtained and subsequently fixed in the original problem, thus resulting in a mp-QP problem. This problem can be solved using the algorithm presented in [213], which results in an initial partitioning of the parameter space and provides a parametric upper bound to the solution. The upper bound for the remaining part of the parameter space, which has not yet been explored is set to infinity.

https://doi.org/10.1515/9783110760316-008

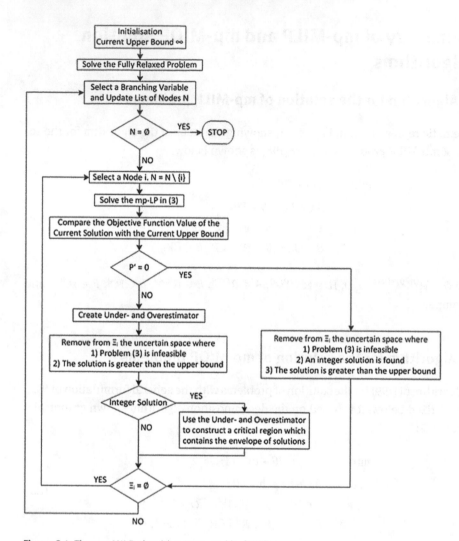

Figure C.1: The mp-MILP algorithm proposed by [198].

Step 1
A candidate solution for the binary variables is found by considering parameters as optimization variables and solving the resulting MIQP problem.

Step 2
Create an affine outer approximation by employing McCormick relaxations [280] for each bilinear or quadratic term in the constraints. Since the nonlinearities in the constraints only arise from comparison procedures, these relaxations are calculated during the comparison procedure.

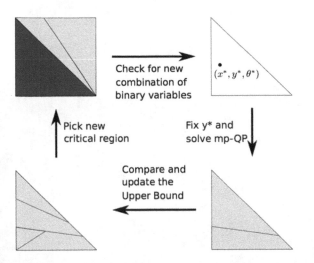

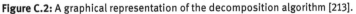

Figure C.2: A graphical representation of the decomposition algorithm [213].

Step 3
The candidate solution of the binary variables is substituted into the initial problem, thus resulting in a mp-QP. This mp-QP problem can be solved using mp-QP algorithms by [213].

Step 4
This and all subsequent steps have to be performed for each critical region. Compare solution with the current upper bound. Here, the explicit solution of the problem is considered, and thus two new critical regions are created.

Step 5
Calculate appropriate relaxations in order to create the outer approximation for the next iteration.

Step 6
The original inequalities from the current critical region are re-introduced to each newly formed critical region, while the relaxations used before are removed. The newly formed critical regions are returned to Step 1 thus resuming the iteration.

Termination
The algorithm terminates as soon as problem in Step 1 is infeasible for all critical regions.

Bibliography

[1]   A. Migdalas. Bilevel programming in traffic planning: Models, methods and challenge. J. Global Optim., 7(4):381–405, 1995.

[2]   L. N. Vicente and P. H. Calamai. Bilevel and multilevel programming: A bibliography review. J. Global Optim., 5(3):291–306, 1994.

[3]   A. Migdalas, P. M. Pardalos, and P. Vrbrand. Multilevel Optimization: Algorithms and Applications. Springer 1st edition, 2012.

[4]   A. Sinha, P. Malo, and K. Deb. A review on bilevel optimization: From classical to evolutionary approaches and applications. IEEE Trans. Evol. Comput., 22:276–295, 2018.

[5]   D. Boyce and L. G. Mattsson. Modeling residential location choice in relation to housing location and road tolls on congested urban highway networks. Transp. Res., Part B, Methodol., 33(8):581–591, 1999.

[6]   H. Yang and S. Yagar. Traffic assignment and signal control in saturated road networks. Transp. Res., Part A Policy Pract., 29(2):125–139, 1995.

[7]   M. L. Tam and W. H. K. Lam. Balance of car ownership under user demand and road network supply conditions – case study in hong kong. J. Urban Plann. Dev., 130(1):24–36, 2004.

[8]   Y. Gao, G. Q. Zhang, J. Lu, and H. M. Wee. Particle swarm optimization for bi-level pricing problems in supply chains. J. Global Optim., 51(2):245–254, 2011.

[9]   D. Miljkovic. Privatizing state farms in Yugoslavia. J. Policy Model., 24(2):169–179, 2002.

[10]   M. J. Robbins and B. J. Lunday. A bilevel formulation of the pediatric vaccine pricing problem. European J. Oper. Res., 248(2):634–645, 2016.

[11]   M. A. Amouzegar and K. Moshirvaziri. Determining optimal pollution control policies: An application of bilevel programming. European J. Oper. Res., 119(1):100–120, 1999.

[12]   C. A. Floudas, Z. H. Gumus, and M. G. Ierapetritou. Global optimization in design under uncertainty: Feasibility test and flexibility index problems. Ind. Eng. Chem. Res., 40(20):4267–4282, 2001.

[13]   M. G. Ierapetritou and E. N. Pistikopoulos. Batch plant design and operations under uncertainty. Ind. Eng. Chem. Res., 35(3):772–787, 1996.

[14]   J. H. Ryu, V. Dua, and E. N. Pistikopoulos. A bilevel programming framework for enterprise-wide process networks under uncertainty. Comput. Chem. Eng., 28(6–7):1121–1129, 2004.

[15]   D. D. Brengel and W. D. Seider. Coordinated design and control optimization of nonlinear processes. Comput. Chem. Eng., 16(9):861–886, 1992.

[16]   M. L. Luyben and C. A. Floudas. Analyzing the interaction of design and control. 1. A multiobjective framework and application to binary distillation synthesis. Comput. Chem. Eng., 18(10):933–969, 1994.

[17]   M. L. Luyben and C. A. Floudas. Analyzing the interaction of design and control. 2. Reactor separator recycle system. Comput. Chem. Eng., 18(10):971–994, 1994.

[18]   P. Tanartkit and L. T. Biegler. A nested, simultaneous approach for dynamic optimization problems. 1. Comput. Chem. Eng., 20(6–7):735–741, 1996.

[19]   H. I. Calvete, C. Galé, and M.-J. Oliveros. Bilevel model for productiondistribution planning solved by using ant colony optimization. Comput. Oper. Res., 38(1):320–327, 2010.

[20]   I. Grossmann. Enterprise-wide optimization: A new frontier in process systems engineering. AIChE J., 51(7):1846–1857, 2005.

[21]   P. Seferlis and N. F. Giannelos. A two-layered optimisation-based control strategy for multi-echelon supply chain networks. Comput. Chem. Eng., 28(5):799–809, 2004.

[22]   A. Mitsos, G. M. Bollas, and P. I. Barton. Bilevel optimization formulation for parameter estimation in liquid-liquid phase equilibrium problems. Chem. Eng. Sci., 64(3):548–559, 2009.

https://doi.org/10.1515/9783110760316-009

[23] I. E. Grossmann. Challenges in the new millennium: Product discovery and design, enterprise and supply chain optimization, global life cycle assessment. Comput. Chem. Eng., 29(1):29–39, 2004.

[24] S. S. Erenguc, N. C. Simpson, and A. J. Vakharia. Integrated production/distribution planning in supply chains: An invited review. European J. Oper. Res., 115(2):219–236, 1999.

[25] C. J. Vidal and M. Goetschalckx. Strategic production-distribution models: A critical review with emphasis on global supply chain models. European J. Oper. Res., 98(1):1–18, 1997.

[26] A. Gupta and C. D. Maranas. A two-stage modeling and solution framework for multisite midterm planning under demand uncertainty. Ind. Eng. Chem. Res., 39(10):3799–3813, 2000.

[27] R. Sousa, N. Shah, and L. G. Papageorgiou. Supply chain design and multilevel planning – an industrial case. Comput. Chem. Eng., 32(11):2643–2663, 2008.

[28] J. Y. Jung, G. Blau, J. F. Pekny, G. V. Reklaitis, and D. Eversdyk. A simulation based optimization approach to supply chain management under demand uncertainty. Comput. Chem. Eng., 28(10):2087–2106, 2004.

[29] E. Roghanian, S. J. Sadjadi, and M. B. Aryanezhad. A probabilistic bi-level linear multi-objective programming problem to supply chain planning. Appl. Math. Comput., 188(1):786–800, 2007.

[30] R. J. Kuo and Y. S. Han. A hybrid of genetic algorithm and particle swarm optimization for solving bi-level linear programming problem – a case study on supply chain model. Appl. Math. Model., 35(8):3905–3917, 2011.

[31] D. Ivanov, B. Sokolov, and A. Pavlov. Dual problem formulation and its application to optimal redesign of an integrated production-distribution network with structure dynamics and ripple effect considerations. Int. J. Prod. Res., 51(18):5386–5403, 2013.

[32] M. Y. Jaber and I. H. Osman. Coordinating a two-level supply chain with delay in payments and profit sharing. Comput. Ind. Eng., 50(4):385–400, 2006. ISSN 0360-8352.

[33] J.-H. Ryu, V. Dua, and E. N. Pistikopoulos. Proactive scheduling under uncertainty: A parametric optimization approach. Ind. Eng. Chem. Res., 46(24):8044–8049, 2007.

[34] V. Sakizlis, N. M. P. Kakalis, V. Dua, J. D. Perkins, and E. N. Pistikopoulos. Design of robust model-based controllers via parametric programming. Automatica, 40(2):189–201, 2004.

[35] S. Subrahmanyam, J. F. Pekny, and G. V. Reklaitis. Design of batch chemical plants under market uncertainty. Ind. Eng. Chem. Res., 33(11):2688–2701, 1994.

[36] S. B. Petkov and C. D. Maranas. Multiperiod planning and scheduling of multiproduct batch plants under demand uncertainty. Ind. Eng. Chem. Res., 36(11):4864–4881, 1997.

[37] M. D. Mesarović, D. Macko, and Y. Takahara. Theory of hierarchical, multilevel, systems. Academic Press, New York, 1970.

[38] R. Scattolini. Architectures for distributed and hierarchical model predictive control – a review. J. Process Control, 19(5):723–731, 2009.

[39] P. D. Christofides, R. Scattolini, D. Munoz de la Pena, and J. Liu. Distributed model predictive control: A tutorial review and future research directions. Comput. Chem. Eng., 51:21–41, 2013.

[40] N. Alguacil, A. Delgadillo, and J. M. Arroyo. A trilevel programming approach for electric grid defense planning. Comput. Oper. Res., 41(1):282–290, 2014.

[41] G. Brown, M. Carlyle, J. Salmeron, and K. Wood. Defending critical infrastructure. Interfaces, 36(6):530–544, 2006.

[42] Y. Yao, T. Edmunds, D. Papageorgiou, and R. Alvarez. Trilevel optimization in power network defense. IEEE Trans. Syst. Man Cybern., Part C Appl. Rev., 37(4):712–718, 2007.

[43] X. Xu, Z. Meng, and R. Shen. A tri-level programming model based on conditional value-at-risk for three-stage supply chain management. Comput. Ind. Eng., 66(2):470–475, 2013.

[44] A. Street, A. Moreira, and J. M. Arroyo. Energy and reserve scheduling under a joint generation and transmission security criterion: An adjustable robust optimization approach. IEEE Trans. Power Syst., 29(1):3–14, 2014.

[45] B. Chen, J. Wang, L. Wang, Y. He, and Z. Wang. Robust optimization for transmission expansion planning: Minimax cost vs. minimax regret. IEEE Trans. Power Syst., 29(6):3069–3077, 2014.

[46] A. Moreira, A. Street, and J. M. Arroyo. An adjustable robust optimization approach for contingency-constrained transmission expansion planning. IEEE Trans. Power Syst., 30(4):2013–2022, 2015.

[47] C. Ning and F. You. Data-driven adaptive nested robust optimization: General modeling framework and efficient computational algorithm for decision making under uncertainty. AIChE J., 63(9):3790–3817, 2017.

[48] S. Avraamidou and E. N. Pistikopoulos. A multiparametric mixed-integer bi-level optimization strategy for supply chain planning under demand uncertainty. IFAC-PapersOnLine, 50(1):10178–10183, 2017.

[49] Y. K. Kim, K. Park, and J. Ko. A symbiotic evolutionary algorithm for the integration of process planning and job shop scheduling. Comput. Oper. Res., 30(8):1151–1171, 2003.

[50] C. Moon, J. Kim, and S. Hur. Integrated process planning and scheduling with minimizing total tardiness in multi-plants supply chain. Comput. Ind. Eng., 43(1–2):331–349, 2002.

[51] F. You and I. E. Grossmann. Design of responsive supply chains under demand uncertainty. Comput. Chem. Eng., 32(12):3090–3111, 2008.

[52] M. E. Dogan and I. E. Grossmann. A decomposition method for the simultaneous planning and scheduling of single-stage continuous multiproduct plants. Ind. Eng. Chem. Res., 45(1):299–315, 2006.

[53] Z. Li and M. G. Ierapetritou. Integrated production planning and scheduling using a decomposition framework. Chem. Eng. Sci., 64(16):3585–3597, 2009.

[54] S. B. Gershwin. Hierarchical flow control: a framework for scheduling and planning discrete events in manufacturing systems. Proc. IEEE, 77(1):195–209, 1989.

[55] S. Bose and J. F. Pekny. A model predictive framework for planning and scheduling problems: a case study of consumer goods supply chain. Comput. Chem. Eng., 24(2–7):329–335, 2000.

[56] J. M. Pinto, M. Joly, and L. F. L. Moro. Planning and scheduling models for refinery operations. Comput. Chem. Eng., 24(9):2259–2276, 2000.

[57] Y. Hae Lee, C. S. Jeong, and C. Moon. Advanced planning and scheduling with outsourcing in manufacturing supply chain. Comput. Ind. Eng., 43(1):351–374, 2002.

[58] H. von Stackelberg, D. Bazin, R. Hill, and L. Urch. Market Structure and Equilibrium. Springer, Berlin Heidelberg, 1934.

[59] J. Bracken and J. McGill. Mathematical programs with optimization problems in the constraints. Oper. Res., 44:21–37, 1973.

[60] W. Candler and R. Townsley. A linear 2-level programming problem. Comput. Oper. Res., 9(1):59–76, 1982.

[61] J. F. Bard. An algorithm for solving the general bilevel programming problem. Math. Oper. Res., 8(2):260–272, 1983.

[62] H. Tuy, A. Migdalas, and P. Värbrand. A global optimization approach for the linear two-level program. J. Global Optim., 3(1):1–23, 1993.

[63] W. F. Bialas and M. H. Karwan. 2-level linear-programming. Manage. Sci., 30(8):1004–1020, 1984.

[64] C. Shi, J. Lu, and G. Zhang. An extended kth-best approach for linear bilevel programming. Appl. Math. Comput., 164(3):843–855, 2005.

[65] J. F. Bard and J. E. Falk. An explicit solution to the multilevel programming problem. Comput. Oper. Res., 9(1):77–100, 1982.

[66] J. Fortuny-Amat and B. McCarl. A representation and economic interpretation of a two-level programming problem. J. Oper. Res. Soc., 32:783–792, 1981.

[67] J. J. Judice and A. Faustino. The solution of the linear bilevel programming problem by using the linear complementarity problem. Investig. Oper., 8:77–95, 1988.

[68] J. J. Judice and A. Faustino. A sequential lcp method for bilevel linear programming. Ann. Oper. Res., 34:89–106, 1992.

[69] V. Visweswaran, C. A. Floudas, M. G. Ierapetritou, and E. N. Pistikopoulos. A decomposition-based global optimization approach for solving bilevel linear and quadratic programs. In State of the Art in Global Optimization, pages 139–162, #7, 1996.

[70] J. F. Bard and J. T. Moore. A branch and bound algorithm for the bilevel programming problem. SIAM J. Sci. Statist. Comput., 11(2):281–292, 1990.

[71] J. J. Judice and A. Faustino. The linear-quadratic bilevel programming problem. INFOR Inf. Syst. Oper. Res., 32(2):87–98, 1994.

[72] L. Vicente, G. Savard, and J. Judice. Descent approaches for quadratic bilevel programming. J. Optim. Theory Appl., 81(2):379–399, 1994.

[73] J. F. Bard. Convex 2-level optimization. Math. Program., 40(1):15–27, 1988.

[74] F. A. Al-Khayyal, R. Horst, and P. M. Pardalos. Global optimization of concave functions subject to quadratic constraints: an application in nonlinear bilevel programming. Ann. Oper. Res., 34(1):125–147, 1992.

[75] T. A. Edmunds and J. F. Bard. Algorithms for nonlinear bilevel mathematical programs. IEEE Trans. Syst. Man Cybern., 21(1):83–89, 1991.

[76] E. Aiyoshi and K. Shimizu. Hierarchical decentralized systems and its new solution by a barrier method. IEEE Trans. Syst. Man Cybern., 11(6):444–449, 1981.

[77] J. F. Bard. An investigation of the linear 3 level programming problem. IEEE Trans. Syst. Man Cybern., 14(5):711–717, 1984.

[78] K. H. Sahin and A. R. Ciric. A dual temperature simulated annealing approach for solving bilevel programming problems. Comput. Chem. Eng., 23(1):11–25, 1998.

[79] P. Marcotte, G. Savard, and D. L. Zhu. A trust region algorithm for nonlinear bilevel programming. Oper. Res. Lett., 29(4):171–179, 2001.

[80] B. Colson, P. Marcotte, and G. Savard. A trust-region method for nonlinear bilevel programming: Algorithm and computational experience. Comput. Optim. Appl., 30(3):211–227, 2005.

[81] Y. F. Yin. Genetic-algorithms-based approach for bilevel programming models. J. Transp. Eng., 126(2):115–120, 2000.

[82] S. Dempe and A. B. Zemkoho. The bilevel programming problem: reformulations, constraint qualifications and optimality conditions. Math. Program., 138(1):447–473, 2013.

[83] S. Dempe and A. B. Zemkoho. Kkt reformulation and necessary conditions for optimality in nonsmooth bilevel optimization. SIAM J. Optim., 24(4):1639–1669, 2014.

[84] A. Tsoukalas, B. Rustem, and E. N. Pistikopoulos. A global optimization algorithm for generalized semi-infinite, continuous minimax with coupled constraints and bi-level problems. J. Global Optim., 44(2):235–250, 2009.

[85] S. Dempe and S. Franke. On the solution of convex bilevel optimization problems. Comput. Optim. Appl., 63:685–-703 (2016).

[86] P. Hansen, B. Jaumard, and G. Savard. New branch-and-bound rules for linear bilevel programming. SIAM J. Sci. Statist. Comput., 13(5):1194–1217, 1992.

[87] A. Mitsos, P. Lemonidis, and P. I. Barton. Global solution of bilevel programs with a nonconvex inner program. J. Global Optim., 42(4):475–513, 2008.

[88] P.-M. Kleniati and C. S. Adjiman. Branch-and-sandwich: a deterministic global optimization algorithm for optimistic bilevel programming problems. part i: Theoretical development. J. Global Optim., 60(3):425–458, 2014.

[89] P.-M. Kleniati and C. S. Adjiman. Branch-and-sandwich: a deterministic global optimization algorithm for optimistic bilevel programming problems. part ii: Convergence analysis and numerical results. J. Global Optim., 60(3):459–481, 2014.

[90]   Z. H. Gumus and C. A. Floudas. Global optimization of mixed-integer bilevel programming problems. Comput. Manag. Sci., 2:181–212, 2005.

[91]   S. Avraamidou and E. N. Pistikopoulos. A multi-parametric optimization approach for bilevel mixed-integer linear and quadratic programming problems. Comput. Chem. Eng., 125:98–113, 2019.

[92]   X. Deng. Complexity issues in bilevel linear programming. In Multilevel Optimization: Algorithms and Applications, pages 149–164, 1998.

[93]   G. K. Saharidis and M. G. Ierapetritou. Resolution method for mixed integer bi-level linear problems based on decomposition technique. J. Global Optim., 44(1):29–51, 2009.

[94]   A. Mitsos. Global solution of nonlinear mixed-integer bilevel programs. J. Global Optim., 47(4):557–582, 2010.

[95]   Z. H. Gumus and C. A. Floudas. Global optimization of nonlinear bilevel programming problems. J. Global Optim., 20(1):1–31, 2001.

[96]   X. Zhu and P. Guo. Approaches to four types of bilevel programming problems with nonconvex nonsmooth lower level programs and their applications to newsvendor problems. Math. Methods Oper. Res., 86:255–-275 (2017).

[97]   U. P. Wen and Y. H. Yang. Algorithms for solving the mixed integer 2-level linear-programming problem. Comput. Oper. Res., 17(2):133–142, 1990.

[98]   U. P. Wen and A. D. Huang. A simple tabu search method to solve the mixed-integer linear bilevel programming problem. European J. Oper. Res., 88(3):563–571, 1996.

[99]   N. P. Faisca, V. Dua, B. Rustem, P. M. Saraiva, and E. N. Pistikopoulos. Parametric global optimisation for bilevel programming. J. Global Optim., 38(4):609–623, 2007.

[100] M. Caramia and R. Mari. A decomposition approach to solve a bilevel capacitated facility location problem with equity constraints. Optim. Lett., 10(5):997–1019, 2016.

[101] P. Fontaine and S. Minner. Benders decomposition for discrete-continuous linear bilevel problems with application to traffic network design. Transp. Res., Part B, Methodol., 70:163–172, 2014.

[102] L. Vicente, G. Savard, and J. Judice. Discrete linear bilevel programming problem. J. Optim. Theory Appl., 89(3):597–614, 1996.

[103] J. F. Bard and J. T. Moore. An algorithm for the discrete bilevel programming problem. Naval Res. Logist., 39(3):419–435, 1992.

[104] S. Dempe. Discrete bilevel optimization problems. Technical report, 2001.

[105] S. T. DeNegre and T. K. Ralphs. A branch-and-cut algorithm for integer bilevel linear programs. Oper. Res. Cyber-Infrastruct., 47:65–78, 2009.

[106] I. Nishizaki and M. Sakawa. Computational methods through genetic algorithms for obtaining stackelberg solutions to two-level integer programming problems. Cybernet. Systems, 36(6):565–579, 2005.

[107] S. D. Handoko, L. H. Chuin, A. Gupta, O. Y. Soon, H. C. Kim, and T. P. Siew. Solving multi-vehicle profitable tour problem via knowledge adoption in evolutionary bi-level programming. In 2015 IEEE Congress on Evolutionary Computation (CEC), pages 2713–2720, 2015.

[108] S. Dempe and F. M. Kue. Solving discrete linear bilevel optimization problems using the optimal value reformulation. J. Global Optim., 68:255–-277 (2017).

[109] S. Dempe. Discrete bilevel optimization problems. Universitaet Leipzig, Wirtschaftswissenschaftliche Fakultaet, Institut für Informatik, 1996.

[110] S. Dempe, K. Richter, T. B. Freiberg, and T. Chemnitz. Bilevel programming with knapsack constraints. CEJOR Cent. Eur. J. Oper. Res., 8(2):93–-107, 2000.

[111] M. Koppe, M. Queyranne, and C. T. Ryan. Parametric integer programming algorithm for bilevel mixed integer programs. J. Optim. Theory Appl., 146(1):137–150, 2010.

[112]   J. T. Moore and J. F. Bard. The mixed integer linear bilevel programming problem. Oper. Res., 38(5):911–921, 1990.

[113]   S. Dempe, V. Kalashnikov, and R. Z. Rios-Mercado. Discrete bilevel programming: Application to a natural gas cash-out problem. European J. Oper. Res., 166(2):469–488, 2005.

[114]   P. Xu and L. Z. Wang. An exact algorithm for the bilevel mixed integer linear programming problem under three simplifying assumptions. Comput. Oper. Res., 41:309–318, 2014.

[115]   P. Xu. Three essays on bilevel optimization algorithms and applications. Ph. D. thesis, Iowa State University, 2012.

[116]   A. Rahmani and S. A. MirHassani. Lagrangian relaxation-based algorithm for bi-level problems. Optim. Methods Softw., 30(1):1–14, 2015.

[117]   D. Yue and F. You. Projection-based reformulation and decomposition algorithm for a class of mixed-integer bilevel linear programs. Comput.-Aided Chem. Eng., 38:481–486, 2016.

[118]   B. Zeng and Y. An. Solving bilevel mixed integer program by reformulations and decomposition. Optimization (Online), 1–34, 2014.

[119]   P.-L. Poirion, S. Toubaline, C. D'Ambrosio, and L. Leo. Bilevel mixed-integer linear programs and the zero forcing set. Optimization (Online), 1–15, 2015.

[120]   M. Fischetti, I. Ljubic, M. Monaci, and M. Sinnl. A new general-purpose algorithm for mixed-integer bilevel linear programs. Oper. Res., 65(6):1615–1637, 2017.

[121]   M. Fischetti, I. Ljubic, M. Monaci, and M. Sinnl. Intersection cuts for bilevel optimization. In Q. Louveaux and M. Skutella, editors, Integer Programming and Combinatorial Optimization, pages 77–88. Springer, 2016.

[122]   M. Fischetti, I. Ljubić, M. Monaci, and M. Sinnl. On the use of intersection cuts for bilevel optimization. Math. Program., 172(1):77–103, 2018.

[123]   T. A. Edmunds and A. J. Bard. An algorithm for the mixed-integer nonlinear bilevel programming problem. Ann. Oper. Res., 34:149–162, 1992.

[124]   R. H. Jan and M. S. Chern. Nonlinear integer bilevel programming. European J. Oper. Res., 72(3):574–587, 1994.

[125]   O. E. Emam. A fuzzy approach for bi-level integer non-linear programming problem. Appl. Math. Comput., 172(1):62–71, 2006.

[126]   L. Hecheng and W. Yuping. Exponential distribution-based genetic algorithm for solving mixed-integer bilevel programming problems. J. Syst. Eng. Electron., 19(6):1157–1164, 2008.

[127]   H. C. Li and Y. P. Wang. Exponential distribution-based genetic algorithm for solving mixed-integer bilevel programming problems. J. Syst. Eng. Electron., 19(6):1157–1164, 2008.

[128]   J. M. Arroyo and F. J. Fernandez. A genetic algorithm approach for the analysis of electric grid interdiction with line switching. In 15th International Conference on Intelligent System Applications to Power Systems, pages 1–6, 2009.

[129]   L. F. Dominguez and E. N. Pistikopoulos. Multiparametric programming based algorithms for pure integer and mixed-integer bilevel programming problems. Comput. Chem. Eng., 34(12):2097–2106, 2010.

[130]   P. M. Kleniati and C. S. Adjiman. A generalization of the branch-and-sandwich algorithm: From continuous to mixed-integer nonlinear bilevel problems. Comput. Chem. Eng., 72:373–386, 2015.

[131]   L. Lozano and J. C. Smith. A value-function-based exact approach for the bilevel mixed-integer programming problem. Oper. Res., 65(3):768–786, 2017.

[132]   B. Beykal, S. Avraamidou, I. PE Pistikopoulos, M. Onel, and E. N. P. Domino. Data-driven optimization of bi-level mixed-integer nonlinear problems. J. Global Optim., 78:1–36, 2020.

[133]   C. Blair. The computational complexity of multi-level linear programs. Ann. Oper. Res., 34(1):13–19, 1992.

[134]   S. Dempe, B. S. Mordukhovich, and A. B. Zemkoho. Necessary optimality conditions in pessimistic bilevel programming. Optimization, 63(4):505–533, 2014.

[135] D. Huppmann and J. Egerer. National-strategic investment in european power transmission capacity. European J. Oper. Res., 247(1):191–203, 2015.

[136] U.-P. Wen and W. F. Bialas. The hybrid algorithm for solving the three-level linear programming problem. Comput. Oper. Res., 13(4):367–377, 1986.

[137] Y.-J. Lai. Hierarchical optimization: A satisfactory solution. Fuzzy Sets and Systems, 77(3):321–335, 1996.

[138] S. Pramanik and T. K. Roy. Fuzzy goal programming approach to multilevel programming problems. European J. Oper. Res., 176(2):1151–1166, 2007.

[139] M. Sakawa, I. Nishizaki, and Y. Uemura. Interactive fuzzy programming for multilevel linear programming problems. Comput. Math. Appl., 36(2):71–86, 1998.

[140] H.-S. Shih, Y.-J. Lai, and E. S. Lee. Fuzzy approach for multi-level programming problems. Comput. Oper. Res., 23(1):73–91, 1996.

[141] D. J. White. Penalty function approach to linear trilevel programming. J. Optim. Theory Appl., 93(1):183–197, 1997.

[142] M. Sakawa and T. Matsui. Interactive fuzzy stochastic multi-level 0-1 programming using tabu search and probability maximization. Expert Syst. Appl., 41(6):2957–2963, 2014.

[143] M. Sakawa, I. Nishizaki, and M. Hitaka. Interactive fuzzy programming for multi-level 0-1 programming problems through genetic algorithms. European J. Oper. Res., 114(3):580–588, 1999.

[144] N. P. Faisca, P. M. Saraiva, B. Rustem, and E. N. Pistikopoulos. A multi-parametric programming approach for multilevel hierarchical and decentralised optimisation problems. Comput. Manag. Sci., 6:377–397, 2009.

[145] J. Han, G. Zhang, Y. Hu, and J. Lu. A solution to bi/tri-level programming problems using particle swarm optimization. Inform. Sci., 370–371:519–537, 2016.

[146] A. T. Woldemariam and S. M. Kassa. Systematic evolutionary algorithm for general multilevel Stackelberg problems with bounded decision variables (seamsp). Ann. Oper. Res., 229:771–790 (2015).

[147] A. M. Kassa and S. M. Kassa. A branch-and-bound multi-parametric programming approach for non-convex multilevel optimization with polyhedral constraints. J. Global Optim., 64(4):745–764, 2016.

[148] C. Shi, G. Zhang, and J. Lu. The kth-best approach for linear bilevel multi-follower programming. J. Global Optim., 33(4):563–578, 2005.

[149] C. Shi, H. Zhou, J. Lu, G. Zhang, and Z. Zhang. The kth-best approach for linear bilevel multifollower programming with partial shared variables among followers. Appl. Math. Comput., 188(2):1686–1698, 2007.

[150] J. Lu, C. Shi, and G. Zhang. On bilevel multi-follower decision making: General framework and solutions. Inform. Sci., 176(11):1607–1627, 2006.

[151] J. Lu, C. Shi, G. Zhang, and T. Dillon. Model and extended Kuhn–Tucker approach for bilevel multi-follower decision making in a referential-uncooperative situation. J. Global Optim., 38(4):597–608, 2007.

[152] G. Bollas, P. I. Barton, and A. Mitsos. Bilevel optimization formulation for parameter estimation in vapor–liquid(–liquid) phase equilibrium problems. Chem. Eng. Sci., 64:1768–1783, 2009.

[153] N. Faísca, P. Saraiva, B. Rustem, and E. Pistikopoulos. A multi-parametric programming approach for multilevel hierarchical and decentralised optimisation problems. Comput. Manag. Sci., 6(4):377–397, 2009. ISSN 1619-697X.

[154] A. Sinha, P. Malo, A. Frantsev, and K. Deb. Finding optimal strategies in a multi-period multi-leader-follower Stackelberg game using an evolutionary algorithm. Comput. Oper. Res., 41(1):374–385, 2014.

[155] T. Gal and J. Nedoma. Multiparametric linear programming. Manage. Sci., 18(7):406–422, 1972.

[156] A. Bemporad, F. Borrelli, and M. Morari. The explicit solution of constrained LP-based receding horizon control. In Proceedings of the 39th IEEE Conference on Decision and Control, 2000, pages 632–637, #1, 2000.

[157] A. Bemporad, F. Borrelli, and M. Morari. Model predictive control based on linear programming – the explicit solution. IEEE Trans. Automat. Control, 47(12):1974–1985, 2002.

[158] F. Borrelli, A. Bemporad, and M. Morari. Geometric Algorithm for Multiparametric Linear Programming. J. Optim. Theory Appl., 118(3):515–540, 2003. ISSN 0022-3239.

[159] P. Tøndel, T. A. Johansen, and A. Bemporad. An algorithm for multi-parametric quadratic programming and explicit MPC solutions. Automatica, 39(3):489–497, 2003. ISSN 0005-1098.

[160] J. Spjotvold, E. C. Kerrigan, C. N. Jones, P. Tondel, and T. A. Johansen. On the facet-to-facet property of solutions to convex parametric quadratic programs. In Mathematical Theory of Networks and Systems, Kyoto, Japan, 2006.

[161] J. Spjøtvold, E. C. Kerrigan, C. N. Jones, P. Tøndel, and T. A. Johansen. On the facet-to-facet property of solutions to convex parametric quadratic programs. Automatica, 42(12):2209–2214, 2006. ISSN 0005-1098.

[162] C. N. Jones, M. Barić, and M. Morari. Multiparametric Linear Programming with Applications to Control. Eur. J. Control, 13(2–3):152–170, 2007.

[163] J. Spjotvold, P. Tondel, and T. A. Johansen. A method for obtaining continuous solutions to multiparametric linear programs. IFAC Proc. Vol., 38(1):253–258, 2005.

[164] S. Olaru and D. Dumur. On the continuity and complexity of control laws based on multiparametric linear programs. In 45th. IEEE Conference on Decision and Control, pages 5465–5470, 2006.

[165] C. N. Jones, E. C. Kerrigan, and J. M. Maciejowski. Lexicographic perturbation for multiparametric linear programming with applications to control. Automatica, 43(10):1808–1816, 2007. ISSN 0005-1098.

[166] C. N. Jones, E. C. Kerrigan, and J. M. Maciejowski. Lexicographic perturbation for multiparametric linear programming with applications to control. Automatica, 43(10):1808–1816, 2007.

[167] H. Milan. Multiparametric linear programming: Support set and optimal partition invariancy. European J. Oper. Res., 202(1):25–31, 2010. ISSN 0377-2217.

[168] T. Gal and H. J. Greenberg. Advances in Sensitivity Analysis and Parametric Programming #6. Springer, Boston, MA, 1997. ISBN 978-0-7923-9917-9.

[169] A. Ghaffari Hadigheh and T. Terlaky. Generalized support set invariancy sensitivity analysis in linear optimization. J. Ind. Manag. Optim., 2(1):1–18, 2006.

[170] A. G. Hadigheh and T. Terlaky. Sensitivity analysis in linear optimization: Invariant support set intervals. European J. Oper. Res., 169(3):1158–1175, 2006. ISSN 0377-2217.

[171] A. Ghaffari Hadigheh, K. Mirnia, and T. Terlaky. Active Constraint Set Invariancy Sensitivity Analysis in Linear Optimization. J. Optim. Theory Appl., 133(3):303–315, 2007. ISSN 0022-3239.

[172] H. J. Greenberg. The use of the optimal partition in a linear programming solution for postoptimal analysis. Oper. Res. Lett., 15(4):179–185, 1994. ISSN 0167-6377.

[173] A. B. Berkelaar, K. Roos, and T. Terlaky. The Optimal Set and Optimal Partition Approach to Linear and Quadratic Programming. In T. Gal and H. J. Greenberg, editors, Advances in Sensitivity Analysis and Parametric Programming, pages 159–202 International Series in Operations Research and Management Science. Springer, Boston, MA, 1997. ISBN 978-1-4615-6103-3.

[174] H. J. Greenberg. Simultaneous Primal-Dual Right-Hand-Side Sensitivity Analysis from a Strictly Complementary Solution of a Linear Program. SIAM J. Optim., 10(2):427–442, 2000.

[175] A. V. Fiacco. Sensitivity analysis for nonlinear programming using penalty methods. Math. Program., 10(1):287–311, 1976. ISSN 0025-5610.

[176] A. Bemporad, M. Morari, V. Dua, and E. N. Pistikopoulos. The explicit solution of model predictive control via multiparametric quadratic programming. In Proceedings of the American Control Conference, pages 872–876, #2, 2000.

[177] A. Bemporad, N. A. Bozinis, V. Dua, M. Morari, and E. N. Pistikopoulos. Model predictive control: A multi-parametric programming approach. In European Symposium on Computer Aided Process Engineering-10, pages 301–306, #8. Elsevier, 2000.

[178] E. N. Pistikopoulos, V. Dua, N. A. Bozinis, A. Bemporad, and M. Morari. On-line optimization via off-line parametric optimization tools. Comput. Chem. Eng., 24(2–7):183–188, 2000. ISSN 0098-1354.

[179] A. Bemporad, M. Morari, V. Dua, and E. N. Pistikopoulos. The explicit linear quadratic regulator for constrained systems. Automatica, 38(1):3–20, 2002. ISSN 0005-1098.

[180] M. Baotic. An Efficient Algorithm for Multiparametric Quadratic Programming, 2002.

[181] P. Tøndel, T. A. Johansen, and A. Bemporad. Further results on multiparametric quadratic programming. In Proceedings of the 42nd IEEE Conference on Decision and Control, 2003, pages 3173–3178, #3, 2003.

[182] D. Q. Mayne. Control of Constrained Dynamic Systems. Eur. J. Control, 7(2–3):87–99, 2001.

[183] D. Q. Mayne and S. Rakovic. Optimal control of constrained piecewise affine discrete time systems using reverse transformation. In Proceedings of the 41st IEEE Conference on Decision and Control, 2002, pages 1546–1551, #2, 2002.

[184] M. M. Seron, G. C. Goodwin, and J. A. de Dona. Finitely parameterised implementation of receding horizon control for constrained linear systems. In Proceedings of the 2002 American Control Conference, 2002, pages 4481–4486, #6, 2002.

[185] D. Q. Mayne and S. Raković. Optimal Control of Constrained Piecewise Affine Discrete-Time Systems. Comput. Optim. Appl., 25(1–3):167–191, 2003. ISSN 0926-6003.

[186] A. Gupta, S. Bhartiya, and P. S. V. Nataraj. A novel approach to multiparametric quadratic programming. Automatica, 47(9):2112–2117, 2011.

[187] C. Feller, T. A. Johansen, and S. Olaru. Combinatorial multi-parametric quadratic programming with saturation matrix based pruning. In 2012 IEEE 51st Annual Conference on Decision and Control (CDC), pages 4562–4567, 2012.

[188] C. Feller and T. A. Johansen. Explicit MPC of higher-order linear processes via combinatorial multi-parametric quadratic programming. In 2013 European Control Conference (ECC), pages 536–541, 2013.

[189] C. Feller, T. A. Johansen, and S. Olaru. An improved algorithm for combinatorial multi-parametric quadratic programming. Automatica, 49(5):1370–1376, 2013. ISSN 0005-1098.

[190] M. Herceg, C. N. Jones, M. Kvasnica, and M. Morari. Enumeration-based approach to solving parametric linear complementarity problems. Automatica, 62:243–248, 2015. ISSN 0005-1098.

[191] R. Oberdieck, N. A. Diangelakis, and E. N. Pistikopoulos. Explicit model predictive control: A connected-graph approach. Automatica, 76:103–112, 2017.

[192] C. N. Jones and M. Morari. Multiparametric Linear Complementarity Problems. In 2006 45th IEEE Conference on Decision and Control, pages 5687–5692, 2006.

[193] M. Herceg, C. N. Jones, M. Kvasnica, and M. Morari. Enumeration-based approach to solving parametric linear complementarity problems. Automatica, 62:243–248, 2015.

[194] P. Patrinos and H. Sarimveis. A new algorithm for solving convex parametric quadratic programs based on graphical derivatives of solution mappings. Automatica, 46(9):1405–1418, 2010. ISSN 0005-1098.

[195] S. B. Olaru and D. Dumur. A parameterized polyhedra approach for explicit constrained predictive control. In 43rd IEEE Conference on Decision and Control, 2004. CDC, pages 1580–1585, #2, 2004.

[196] M. Monnigmann and M. Jost. Vertex based calculation of explicit MPC laws. In American Control Conference (ACC), 2012, pages 423–428, 2012.

[197] M. Wittmann-Hohlbein and E. N. Pistikopoulos. A Two-Stage Method for the Approximate Solution of General Multiparametric Mixed-Integer Linear Programming Problems. Ind. Eng. Chem. Res., 51(23):8095–8107, 2012.

[198] R. Oberdieck, M. Wittmann-Hohlbein, and E. N. Pistikopoulos. A branch and bound method for the solution of multiparametric mixed integer linear programming problems. J. Global Optim., 59(2–3):527–543, 2014.

[199] R. Oberdieck and E. N. Pistikopoulos. Explicit hybrid model-predictive control: The exact solution. Automatica, 58:152–159, 2015.

[200] A. Bemporad, F. Borrelli, and M. Morari. Piecewise linear optimal controllers for hybrid systems. In Proceedings of the American Control Conference, pages 1190–1194, #2, 2000.

[201] A. Bemporad, F. Borrelli, and M. Morari. Optimal controllers for hybrid systems: stability and piecewise linear explicit form. In Proceedings of the 39th IEEE Conference on Decision and Control, 2000, pages 1810–1815, #2, 2000.

[202] F. Borelli. Constrained optimal control of linear and hybrid systems. Springer, New York, 2003. ISBN 978-3-540-00257-4.

[203] J. Acevedo and E. N. Pistikopoulos. A Multiparametric Programming Approach for Linear Process Engineering Problems under Uncertainty. Ind. Eng. Chem. Res., 36(3):717–728, 1997.

[204] J. Acevedo and E. N. Pistikopoulos. An algorithm for multiparametric mixed-integer linear programming problems. Oper. Res. Lett., 24(3):139–148, 1999. ISSN 0167-6377.

[205] V. Dua and E. N. Pistikopoulos. An Algorithm for the Solution of Multiparametric Mixed Integer Linear Programming Problems. Ann. Oper. Res., 99(1–4):123–139, 2000. ISSN 0254-5330.

[206] Z. Jia and M. G. Ierapetritou. Uncertainty analysis on the righthand side for MILP problems. AIChE J., 52(7):2486–2495, 2006. ISSN 1547-5905.

[207] A. Crema. A contraction algorithm for the multiparametric integer linear programming problem. European J. Oper. Res., 101(1):130–139, 1997. ISSN 0377-2217.

[208] A. Crema. An algorithm to perform a complete right-hand-side parametrical analysis for a 0–1-integer linear programming problem. European J. Oper. Res., 114(3):569–579, 1999. ISSN 0377-2217.

[209] A. Crema. An algorithm for the multiparametric 0–1-integer linear programming problem relative to the constraint matrix. Oper. Res. Lett., 27(1):13–19, 2000. ISSN 0167-6377.

[210] A. Crema. An algorithm for the multiparametric 0–1-integer linear programming problem relative to the objective function. European J. Oper. Res., 125(1):18–24, 2000. ISSN 0377-2217.

[211] A. Crema. An algorithm to perform a complete parametric analysis relative to the constraint matrix for a 0–1-integer linear program. European J. Oper. Res., 138(3):484–494, 2002. ISSN 0377-2217.

[212] A. Crema. The multiparametric 0–1-integer linear programming problem: A unified approach. European J. Oper. Res., 139(3):511–520, 2002. ISSN 0377-2217.

[213] V. Dua, N. A. Bozinis, and E. N. Pistikopoulos. A multiparametric programming approach for mixed-integer quadratic engineering problems. Comput. Chem. Eng., 26(4–5):715–733, 2002.

[214] D. Axehill, Th. Besselmann, D. M. Raimondo, and M. Morari. Suboptimal Explicit Hybrid MPC via Branch and Bound. In IFAC World Congress, Milano, 2011.

[215] D. Axehill, T. Besselmann, D. M. Raimondo, and M. Morari. A parametric branch and bound approach to suboptimal explicit hybrid MPC. Automatica, 50(1):240–246, 2014. ISSN 0005-1098.

[216] A. Fuchs, D. Axehill, and M. Morari. Lifted Evaluation of mp-MIQP Solutions. IEEE Trans. Automat. Control, 60(12):3328–3331, 2015.

[217] T. Gal. Rim multiparametric linear programming. Manage. Sci., 21(5):567–575, 1975.

[218] P. L. Yuf and M. Zeleny. Linear Multiparametric Programming by Multicriteria Simplex Method. Manage. Sci., 23(2):159–170, 1976.

[219] M. Schechter. Polyhedral functions and multiparametric linear programming. J. Optim. Theory Appl., 53(2):269–280, 1987.

[220] A. Pertsinidis, I. E. Grossmann, and G. J. McRae. Parametric optimization of milp programs and a framework for the parametric optimization of minlps. Comput. Chem. Eng., 22:S205–S212, 1998.

[221] M. Baotic. An algorithm for multiparametric quadratic programming. Technical Report, ETH Zurich, 2002.

[222] C. Filippi. An algorithm for approximate multiparametric linear programming. J. Optim. Theory Appl., 120(1):73–95, 2004.

[223] Z. Li and M. G. Ierapetritou. A New Methodology for the General Multiparametric Mixed-Integer Linear Programming (MILP) Problems. Ind. Eng. Chem. Res., 46(15):5141–5151, 2007.

[224] N. P. Faísca and V. D. Kosmidis. Berç Rustem, and Efstratios N Pistikopoulos. Global optimization of multi-parametric milp problems. J. Global Optim., 45(1):131–151, 2009.

[225] A. Mitsos and P. I. Barton. Parametric mixed-integer 0–1 linear programming: The general case for a single parameter. European J. Oper. Res., 194(3):663–686, 2009.

[226] Z. Li and M. G. Ierapetritou. A method for solving the general parametric linear complementarity problem. Ann. Oper. Res., 181(1):485–501, 2010. ISSN 0254-5330.

[227] M. Wittmann-Hohlbein and E. N. Pistikopoulos. On the global solution of multi-parametric mixed integer linear programming problems. J. Global Optim., 57(1):51–73, 2013. ISSN 0925-5001.

[228] R. Oberdieck, M. Wittmann-Hohlbein, and E. N. Pistikopoulos. A branch and bound method for the solution of multiparametric mixed integer linear programming problems. J. Global Optim., 59(2–3):527–543, 2014. ISSN 0925-5001.

[229] M. Wittmann-Hohlbein and E. N. Pistikopoulos. Approximate solution of mp-milp problems using piecewise affine relaxation of bilinear terms. Comput. Chem. Eng., 61:136–155, 2014.

[230] A. Bemporad. A Multiparametric Quadratic Programming Algorithm with Polyhedral Computations Based on Nonnegative Least Squares. IEEE Trans. Automat. Control, 60(11):2892–2903, 2015.

[231] R. Oberdieck and E. N. Pistikopoulos. Explicit hybrid model-predictive control: The exact solution. Automatica, 58(0):152–159, 2015. ISSN 0005-1098.

[232] V. M. Charitopoulos, L. G. Papageorgiou, and V. Dua. Multi-parametric mixed integer linear programming under global uncertainty. Comput. Chem. Eng., 116:279–295, 2018.

[233] P. Ahmadi-Moshkenani, T. A. Johansen, and S. Olaru. Combinatorial approach towards multi-parametric quadratic programming based on characterizing adjacent critical regions. IEEE Trans. Automat. Control, 63:3221–-3231, 2018.

[234] A. Akbari and P. I. Barton. An improved multi-parametric programming algorithm for flux balance analysis of metabolic networks. J. Optim. Theory Appl., 178:502–-537, 2018.

[235] B. Burnak, J. Katz, and E. N. Pistikopoulos. A space exploration algorithm for multiparametric programming via delaunay triangulation. Optim. Eng., 22:555–-579, 2021.

[236] S. Dempe. Foundations of bi-level programming, 2002.

[237] E. N. Pistikopoulos, V. Dua, and J. H. Ryu. Global optimization of bilevel programming problems via parametric programming. In Frontiers in Global Optimization, pages 457–476, #74, 2003.

[238] E. N. Pistikopoulos, M. C. Georgiadis, and V. Dua. Multi-parametric programming. Volume 1, Theory, algorithms and applications. John Wiley, distributor, Weinheim: Wiley-VCH Chichester, 2007.

[239] N. P. Faisca, V. Dua, P. M. Saraiva, B. Rustem, and E. N. Pistikopoulos. A global parametric programming optimisation strategy for multilevel problems. In 16th European Symposium on Computer Aided Process Engineering and 9th International Symposium on Process Systems Engineering, pages 215–220, #21, 2006.

[240] R. Oberdieck, N. A. Diangelakis, S. Avraamidou, and E. N. Pistikopoulos. On unbounded and binary parameters in multi-parametric programming: applications to mixed-integer bilevel optimization and duality theory. J. Global Optim., 69:587-–606, 2017.

[241] C. A. Floudas. Nonlinear and Mixed-Integer Optimization: Fundamentals and Applications, 1995.

[242] R. Oberdieck, N. A. Diangelakis, M. M. Papathanasiou, I. Nascu, and E. N. Pistikopoulos. Pop – parametric optimization toolbox. Ind. Eng. Chem. Res., 55(33):8979–8991, 2016.

[243] S. Avraamidou, N. A. Diangelakis, and E. N. Pistikopoulos. Mixed integer bilevel optimization through multi-parametric programming. Foundations of Computer Aided Process Operations/Chemical Process Control, in press. https://folk.ntnu.no/skoge/prost/proceedings/focapo-cpc-2017/FOCAPO-CPC%202017%20Contributed%20Papers/73_FOCAPO_Contributed.pdf.

[244] G. Anandalingam. A mathematical programming model of decentralized multi-level systems. J. Oper. Res. Soc., 39(11):1021–1033, 1988.

[245] S. Sinha. A comment on anandalingam (1988). a mathematical programming model of decentralized multi-level systems. j opl res soc 39: 1021–1033. J. Oper. Res. Soc., 52(5):594–596, 2001.

[246] R. Misener and C. A. F. Antigone. Algorithms for continuous / integer global optimization of nonlinear equations. J. Global Optim., 59(2–3):503–526, 2014.

[247] M. Tawarmalani and N. V. Sahinidis. A polyhedral branch-and-cut approach to global optimization. Math. Program., 103(2):225–249, 2005.

[248] A. Dolgui and M.-A. Ould-Louly. A model for supply planning under lead time uncertainty. Int. J. Prod. Econ., 78(2):145–152, 2002.

[249] B. Mansoornejad, E. N. Pistikopoulos, and P. R. Stuart. Scenario-based strategic supply chain design and analysis for the forest biorefinery using an operational supply chain model. Int. J. Prod. Econ., 144(2):618–634, 2013.

[250] M. R. Katebi and M. A. Johnson. Predictive control design for large-scale systems. Automatica, 33(3):421–425, 1997.

[251] S. Avraamidou and E. N. Pistikopoulos. A multi-parametric bi-level optimization strategy for hierarchical model predictive control. In 27th European Symposium on Computer-Aided Process Engineering (ESCAPE-27), pages 1591–1596, 2017.

[252] A. Bemporad, M. Morari, V. Dua, and E. N. Pistikopoulos. Erratum: The explicit linear quadratic regulator for constrained systems. Automatica, 39(10):1845–1846, 2003.

[253] L. El Ghaoui and H. Lebret. Robust solutions to least-squares problems with uncertain data. SIAM J. Matrix Anal. Appl., 18(4):1035–1064, 1997.

[254] L. El Ghaoui, F. Oustry, and H. Lebret. Robust solutions to uncertain semidefinite programs. SIAM J. Optim., 9(1):33–52, 1998.

[255] A. Ben-Tal and A. Nemirovski. Robust convex optimization. Math. Oper. Res., 23(4):769–805, 1998.

[256] A. Ben-Tal and A. Nemirovski. Robust solutions of uncertain linear programs. Oper. Res. Lett., 25(1):1–13, 1999.

[257] A. Ben-Tal and A. Nemirovski. Robust solutions of linear programming problems contaminated with uncertain data. Math. Program., 88(3):411–424, 2000.

[258] D. Bertsimas and M. Sim. The price of robustness. Oper. Res., 52(1):35–53, 2004.

[259] D. Bertsimas and D. B. Brown. Constructing uncertainty sets for robust linear optimization. Oper. Res., 57(6):1483–1495, 2009.

[260] D. Bertsimas, D. A. Iancu, and P. A. Parrilo. A hierarchy of near-optimal policies for multistage adaptive optimization (technical report). IEEE Trans. Automat. Control, 56(12):2809, 2011.

[261] R. Hettich and K. O. Kortanek. Semi-infinite programming: Theory, methods, and applications. SIAM Rev., 35(3):380–429, 1993.

[262] F. Guerra Vazquez, J.-J. Ruckmann, O. Stein, and G. Still. Generalized semi-infinite programming: A tutorial. J. Comput. Appl. Math., 217(2):394–419, 2008. Special Issue: Semi-infinite Programming (SIP).

[263] M. Lopez and G. Still. Semi-infinite programming. European J. Oper. Res., 180(2):491–518, 2007.

[264] A. Ben-Tal, A. Goryashko, E. Guslitzer, and A. Nemirovski. Adjustable robust solutions of uncertain linear programs. Math. Program., 99(2):351–376, 2004.

[265] J. Zhen, D. den Hertog, and M. Sim. Adjustable robust optimization via fourier-motzkin elimination, Optimization (Online), 2017. https://doi.org/10.1287/opre.2017.1714.

[266] G. A. Hanasusanto, D. Kuhn, and W. Wiesemann. K-adaptability in two-stage robust binary programming. Oper. Res., 63(4):877–891, 2015.

[267] D. Bertsimas and A. Georghiou. Design of near optimal decision rules in multistage adaptive mixed-integer optimization. Oper. Res., 63(3):610–627, 2015.

[268] D. Bertsimas and A. Georghiou. Binary decision rules for multistage adaptive mixed-integer optimization. Math. Program., 167, 395–-433, 2018.

[269] B. Zeng and L. Zhao. Solving two-stage robust optimization problems using a column-and-constraint generation method. Oper. Res. Lett., 41:457–461, 2013.

[270] E. N. Pistikopoulos and N. A. Diangelakis. Towards the integration of process design, control and scheduling: Are we getting closer? Comput. Chem. Eng., 91:85–92, 2016. ISSN 0098-1354.

[271] M. Erdirik-Dogan, I. E. Grossmann, and J. Wassick. A bi-level decomposition scheme for the integration of planning and scheduling in parallel multi-product batch reactors. In Computer Aided Chemical Engineering, pages 625–630, #24. Elsevier, 2007.

[272] Z. Lukszo and P. Heijnen. Better design and operation of infrastructures through bi-level decision making. In 2007 IEEE International Conference on Networking, Sensing and Control, pages 181–186. IEEE, 2007.

[273] D. Yue and F. You. Bilevel optimization for design and operations of noncooperative biofuel supply chains. In Chemical Engineering Transactions, pages 1309–1314. Italian Association of Chemical Engineering-AIDIC, 2015.

[274] R. Oberdieck, N. A. Diangelakis, I. Nascu, M. M. Papathanasiou, M. Sun, S. Avraamidou, and E. N. Pistikopoulos. On multi-parametric programming and its applications in process systems engineering. Chem. Eng. Res. Des., 116:61–82, 2016.

[275] E. N. Pistikopoulos, N. A. Diangelakis, R. Oberdieck, M. M. Papathanasiou, I. Nascu, and M. Sun. Paroc – an integrated framework and software platform for the optimisation and advanced model-based control of process systems. Chem. Eng. Sci., 136 (Special Issue):115–138, 2015.

[276] Process Systems Enterprise. gPROMS.

[277] R. S. C. Lambert, P. Rivotti, and E. N. Pistikopoulos. A monte-carlo based model approximation technique for linear model predictive control of nonlinear systems. Comput. Chem. Eng., 54(0):60–67, 2013.

[278] R. S. C. Lambert. Approximation Methodologies for Explicit Model Predictive Control of Complex Systems. PhD thesis, Imperial College, London 2013.

[279] E. N. Pistikopoulos, L. Dominguez, and C. Panos. Konstantinos Kouramas, and Altannar Chinchuluun. Theoretical and algorithmic advances in multi-parametric programming and control. Comput. Manag. Sci., 9(2):183–203, 2012. ISSN 1619-697X.

[280] G. P. McCormick. Computability of global solutions to factorable nonconvex programs: Part I – Convex underestimating problems. Math. Program., 10(1):147–175, 1976.

Index

https://doi.org/10.1515/9783110760316-010

Printed in the USA
CPSIA information can be obtained
at www.ICGtesting.com
JSHW050017170524
63288JS00010B/174

9 783110 760309